QUALITY
FUNCTION
DEPLOYMENT

QUALITY AND RELIABILITY

A Series Edited by

Edward G. Schilling

Center for Quality and Applied Statistics
Rochester Institute of Technology
Rochester, New York

QUALITY FUNCTION DEPLOYMENT

A Practitioner's Approach

James L. Bossert

ASQC Quality Press   Milwaukee, Wisconsin

Marcel Dekker, Inc.   New York • Basel • Hong Kong

Library of Congress Cataloging--in--Publication Data

Bossert, James L.
  Quality function deployment: a practitioner's approach / James L.
Bossert.
     p. cm. ---- (Quality and reliability: 21)
  Includes bibliographical references and index.
  ISBN 0-8247-8378-6
  1. Production management----Quality control. 2. Production
planning. I. Title. II. Series.
  TS156.B665   1991
  658.5'62---dc20   90--43245
                                                                    CIP

This book is printed on acid-free paper.

ASQC Quality Press
310 West Wisconsin Avenue, Milwaukee, Wisconsin 53203

Marcel Dekker, Inc.
270 Madison Avenue, New York, New York 10016

Current printing (last digit):
10 9 8 7 6 5 4 3 2 1

PRINTED IN THE UNITED STATES OF AMERICA

This book is dedicated to

my wife, Nancy,

and my daughters,

Lindsay and Ashley.

They encourage me to excel.

About the Series

The genesis of modern methods of quality and reliability will be found in a simple memo dated May 16, 1924, in which Walter A. Shewhart proposed the control chart for the analysis of inspection data. This led to a broadening of the concept of inspection from emphasis on detection and correction of defective material to control of quality through analysis and prevention of quality problems. Subsequent concern for product performance in the hands of the user stimulated development of the systems and techniques of reliability. Emphasis on the consumer as the ultimate judge of quality serves as the catalyst to bring about the integration of the methodology of quality with that of reliability. Thus, the innovations that came out of the control chart spawned a philosophy of control of quality and reliability that has come to include not only the methodology of the statistical sciences and engineering, but also the use of appropriate management methods together with various motivational procedures in a concerted effort dedicated to quality improvement.

This series is intended to provide a vehicle to foster interaction of the elements of the modern approach to quality, including statistical applications, quality and reliability engineering, management, and motivational aspects. It is a forum in which the subject matter of these various areas can be brought together to allow for effective integration of appropriate techniques. This will promote the true benefit of each, which can be achieved only through their interaction. In this sense, the whole of quality and reliability is greater than the sum of its parts, as each element augments the others.

The contributors to this series have been encouraged to discuss fundamental concepts as well as methodology, technology, and procedures at the leading edge of the discipline. Thus, new concepts are placed in proper perspective in these evolving disciplines.

The series is intended for those in manufacturing, engineering, and marketing and management, as well as the consuming public, all of whom have an interest and stake in the improvement and maintenance of quality and reliability in the products and services that are the lifeblood of the economic system.

The modern approach to quality and reliability concerns excellence: excellence when the product is designed, excellence when the product is made, excellence as the product is used, and excellence throughout its lifetime. But excellence does not result without effort, and products and services of superior quality and reliability require an appropriate combination of statistical, engineering, management, and motivational effort. This effort can be directed for maximum benefit only in light of timely knowledge of approaches and methods that have been developed and are available in these areas of expertise. Within the volumes of this series, the reader will find the means to create, control, correct, and improve quality and reliability in ways that are cost effective, that enhance productivity, and that create a motivational atmosphere that is harmonious and constructive. It is dedicated to that end and to the readers whose study of quality and reliability will lead to greater understanding of their products, their processes, their workplaces, and themselves.

Edward G. Schilling

Preface

This book was written for two types of people: for managers who want to get some idea of what Quality Function Deployment (QFD) is and for the practitioner. Practitioners are the people in your company who are given the task of implementing QFD. This book assumes that they have been to a QFD facilitator's course at GOAL/QPC, ASI, or any of the other organizations that are involved in QFD training. GOAL/QPC and ASI have excellent training programs in the fundamentals. At best, I could only mirror what they do. So in writing this, I hope to give some practical tips on how to proceed. I will pass along things that I have learned in both successes and failures in over two years of implementing QFD exercises.

Managers can use this book to get an idea of what commitments are necessary to obtain success when trying QFD for the first time. Managers today are bombarded with all sorts of tools, techniques, and ideas to try in an effort to make things work more efficiently. Prior to this book, there were only management overviews and numerous articles to get more information on this thing known as QFD. Hopefully, with this book, QFD will be tried and its benefits will reinforce its further use.

James L. Bossert

Acknowledgments

There were many people who influenced me in the writing of this book:

Lynn Hotter, who introduced me to QFD. His patience while I learned the basics and struggled with the first teams was an example for me to follow.

Dick Gruzosky, for his insight into looking at QFD as a system and his encouragement for finding alternative approaches.

Grant Kosten, for his work on the graphics.

Mike Brassard and Bob King, for permission to use some of their materials.

My wife, Nancy, and children, Lindsay and Ashley, for their support, patience, and in Nan's case, her fortitude in deciphering my writing and doing all the typing.

Thanks to all of you.

Contents

QUALITY
FUNCTION
DEPLOYMENT

PART ONE

1

What Is Quality Function Deployment?

When first approached about Quality Function Deployment (QFD), most managers question what it is. Quality Function Deployment, or QFD as it is commonly known, is a process that provides structure to the development cycle. This structure can be likened to the framework of a house. The foundation is customer requirements. The frame consists of the planning matrix, which includes items such as the importance rating, customer-perceived benchmarking, sales point, and scale-up factors. The second floor of the house includes the technical features. The roof is the trade-off of technical features. The walls are the interrelationship matrix between the customer requirements and the technical characteristics. Other parts can be built using things such as new technologies, functions, technical characteristics, processing steps, importance ratings, competitive analysis, and sales points. The components utilized are dependent upon the scope of the project.

The thing that makes QFD unique is that the primary focus is the customer requirements. The process is driven by what the customer wants, not by innovations in technology. Consequently, more effort is involved getting the information necessary for determining what the customer truly wants. This tends

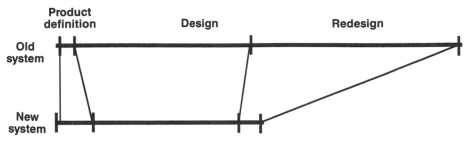

Figure 1
COMPARISON OF OLD AND NEW DESIGN SYSTEMS

to increase the initial planning time in the project definition phase of the development cycle, but it reduces the overall cycle time in bringing a product to market. This is illustrated in Figure 1, which is reproduced with permission from GOAL/QPC. When a product is conceived, the primary focus is on who the customer is, since the customer sets the stage for all the work. What the customer wants will determine whether new technologies are needed, whether simple improvements are possible, or whether a revolutionary concept is required. Success in determining customer requirements is directly related to success in the marketplace. This is critical to the whole process.

Once a product is defined, QFD enables the design phase to focus on the key customer requirements, those elements that are defined as being very important to the customer. By addressing these elements, the design phase is shortened to focus on items that the customer really wants. By concentrating efforts, less time will be spent on redesign and modifications. The savings have been currently estimated as one-third to one-half of the time taken using traditional means. If a new product took eighteen months from concept to market, using QFD could reduce the time to nine to twelve months, with little if any changes to the product once it is in the marketplace. For many companies, this can mean many dollars saved not only in development but also in additional income brought in due to getting out a product that met the customer's needs faster than before.

Another way to think of QFD is to compare the product development cycle to the human body. QFD would be the skeleton, which provides form and structure to the cycle and serves as the framework that ties all activities into a complete package. Without the frame, there would be activity, but it would be vulnerable to outside influences like personnel changes, product redesign, fickle management, and so on. QFD documents all pertinent information to keep everyone on track. When critical decisions are made, they are documented so that all people on the team are not only aware of them but also buy into those decisions.

Once QFD is used, there is a tendency to wonder how things were ever accomplished before. The danger is using QFD as an end to itself. QFD is simply a tool to be utilized where appropriate. Like all tools, there are both proper and improper ways to use it. This book will demonstrate some of the proper ways to effectively use the QFD process. In the following chapters, we will show some ways to get started in the QFD process; how to get customer information from a variety of sources; how to look at the matrices, various tools that can be utilized to get the job done, and potential pitfalls and how to counter them; and some different ways to use the QFD process. The question to ask yourself with any QFD matrix is, "Does this make sense?" If it does, then you are probably using the tool effectively. If not, then re-examine your intent for doing the QFD. There may be another tool that will better fit your needs.

What are the benefits of using QFD? Since QFD is a customer-driven process, it creates a strong focus on the customer. QFD exercises tend to look beyond the usual customer feedback and attempt to define the requirements in a set of basic needs, which are compared to all competitive information available. Therefore, all competitors are evaluated equally both from the customer's perspective and from a technical perspective. Once this information is in hand, then, through a Pareto ranking, the requirements are prioritized, and the manager can then effectively place resources where they can do the most good—on the requirements that are meaningful to the customer and that can be acted upon. An example of this is a plastic part on a car door that was warped by extreme heat when sitting in a parking lot all day. The solution that was found was to use a polymer that was not as sensitive to heat changes over extended periods of time. We could do something about the warping, not about the heat.

Another benefit of QFD is that it structures experience and information into a concise format. In many companies, there is a wealth of information available but not put together in a document. QFD places that information into a structured format that is easy to assimilate. This information contains all necessary rationale for choosing the design, identifying trade-offs, and listing future enhancements. This is important for the times when there are personnel who leave the project and new people are brought on board, as the documentation allows for the swift integration of ideas and progress. QFD is also flexible enough to adapt to new information, since the matrix structure will grow or shrink based on the information received. In essence, QFD produces a living document, one that reacts to input and better defines real needs.

The QFD process is a very robust process. This means that things can be changed in the structure, but when done correctly, the top results do not really change. One QFD exercise involved the development of twenty-five customer requirements. The project leader decided to address the top eight for further development work. He was concerned that if the importance ratings changed, the priority of the requirements would change. A computer program was set up so that all the importance ratings could be changed, holding everything else constant. The result was that the top eight items were always the top eight items, that only the order of occurrence changed. For example, the number one item might move to number three on the priority listing. This goes a long way in relieving concerns that managers may have when going through the QFD exercise.

As people work through the QFD process, a team grows. It is one of the best approaches for developing teamwork, since all decisions are based on consensus and a fair amount of discussion takes place. This discussion allows everyone to explain their

4

views, reach a consensus, and move forward on the project. People feel that all their interests are addressed, not always to their satisfaction, but at least their opinions are heard. This communication at the functional interface is critical. As people see what the larger picture is, individual concerns may not be as critical and some rationalization takes place. The process also identifies what actions need to take place so that all team members see how they fit into the overall project. This solidifies the team membership aspect and encourages teamwork.

When one first looks at a QFD matrix, it may appear to be very confusing and busy, with a lot of detail in each matrix. In truth, there is, but now you can take a step back and look at the overall picture. The matrix enables the creation of a global viewpoint from all the detail, a benefit that is frequently overlooked. Many times we get hung up on the details and forget the original purpose.

In terms of age, QFD is a relatively young technique. Its first use was reported in Japan in the late 60s, at the Kobe Shipyard. From there, the applications were further developed in various industries. In the mid 80s, Dr. Donald Clausing brought information on QFD to Xerox. From there a number of organizations have promoted and taught people in the U.S. how to utilize QFD, the two most prominent organizations being GOAL/QPC and the American Supplier Institute.

In order to truly understand the impact of QFD, one has to look at what happens to a team and an organization when the commitment is made to do QFD. Initially, there is reluctance and skepticism. The thought of another Japanese technique is not necessarily well received. As the team progresses, however, there is a conditional acceptance of what is going on, that there may be some value to QFD. Upon completion of the project, there tend to be three camps: one group that sees it as a good tool with some future potential, the zealots who are converts to the QFD mentality, and the one camp who does not see any value to the technique. Fortunately, the majority of the people fall in the first camp. This is the first step in the defining of a culture change. This is covered in Chapter 2. Figure 2 summarizes the benefits of QFD.

Figure 2
QFD BENEFITS

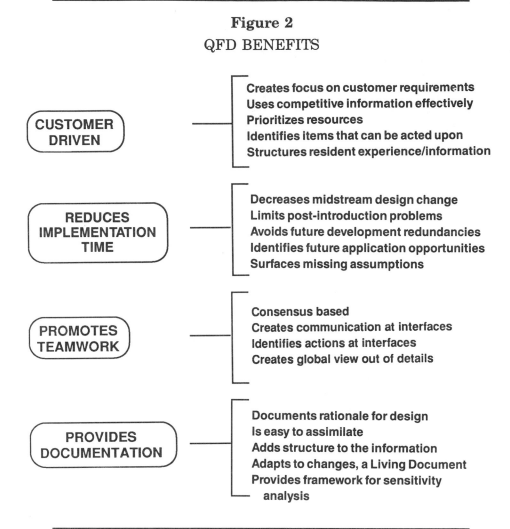

CUSTOMER DRIVEN
- Creates focus on customer requirements
- Uses competitive information effectively
- Prioritizes resources
- Identifies items that can be acted upon
- Structures resident experience/information

REDUCES IMPLEMENTATION TIME
- Decreases midstream design change
- Limits post-introduction problems
- Avoids future development redundancies
- Identifies future application opportunities
- Surfaces missing assumptions

PROMOTES TEAMWORK
- Consensus based
- Creates communication at interfaces
- Identifies actions at interfaces
- Creates global view out of details

PROVIDES DOCUMENTATION
- Documents rationale for design
- Is easy to assimilate
- Adds structure to the information
- Adapts to changes, a Living Document
- Provides framework for sensitivity analysis

When we speak about developing a matrix, we will be utilizing a combination GOAL/QPC call the A-1 matrix and what ASI calls the customer requirements matrix. The matrix looks like that in Figure 3. All other matrices use a similar format. We will discuss the systems approach in Chapter 7.

Figure 3

A BASIC QFD MATRIX SHOWING THE VARIOUS COMPONENTS

QFD PROCESS
CONCEPT
(HOUSE OF QUALITY)

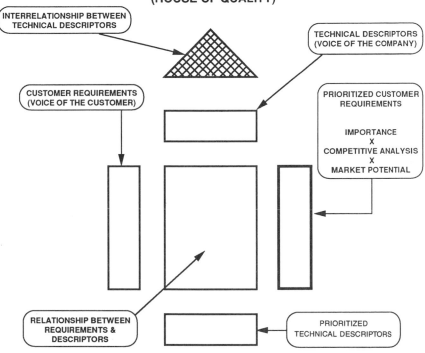

Figure 3 shows a typical starting matrix, which is the foundation of all QFD exercises. Experience has shown that this matrix is also the most difficult one to put together, since it requires a large amount of information from a variety of sources.

On the left side are the customer requirements, what the customer wants in the product. In Chapter 3, we will discuss how to obtain this information as well as how to look beyond it to find the hidden requirements.

The top of the matrix shows the manufacturer's requirements, what the manufacturer does to ensure the consistency of product. These can be items that are measured by the manufacturer or are specified from suppliers.

The right side of the matrix illustrates the planning matrix. This is where the importance rating, the competitive analysis, the target value, the amount of scale up necessary, and the sales point are listed. From this information, a planning weight will be calculated. This planning weight will help the team focus on the items that will yield the greatest potential for success in the marketplace.

The peak of the matrix is the manufacturer's requirements. This is where trade-offs are identified. By identifying these early on, product development people can narrow their development efforts, thus speeding up the development cycle.

The body of the matrix is where the relationships are categorized. This is where customer requirements are "translated" into manufacturer's terms. It is also where interactions between relationships are identified so that the synergistic effect is seen.

The bottom is the prioritized manufacturer's requirements. This identifies the requirements that are the most critical for success as well as the degree of technical difficulty to achieve.

All other matrices will be comprised of these fundamental features. Once the format is understood, all matrices are easily understood.

2

How to Get Started in QFD

One of the first questions a manager will ask is, "What do I need to do?" It is important that prior to starting any QFD exercise a manager take the following steps.

First, the manager must be willing to commit his or her people with respect to time. Too often a QFD exercise will fail because the team cannot get together. A manager must be willing to give the people on the project all the time necessary for them to complete their assignments.

Second, a manager needs to ensure that all team members are aware of the importance of the QFD project. The best people in most departments are also the busiest. Consequently, they will treat the QFD as just another assignment unless some emphasis is placed on its importance to the company. All team members must understand this importance so that they will spend an appropriate amount of time on the exercise.

Third, the scope of the project must be clearly defined and understood by all team members. This helps to avoid the "Why are we here?" questions, which arise wherever the scope is not understood. By not defining the scope, you will add ten to thirty hours to the QFD exercise. This time will be spent on defining the scope and getting team buy-in to that scope.

The final thing that a manager needs to do in the initial stage is to inform all other managers about the QFD project, its scope, and the team members. This helps to eliminate problems when a team member is asked to do more than he or she has time or energy for.

These steps are necessary to set the stage for success in a QFD exercise. The next consideration is what reporting should the manager expect. Experience has shown that if the team leader reports to the manager on a monthly basis, there is enough time to measure progress. In some cases, this reporting could follow a biweekly cycle. The limiting factors here are how frequently the team meets and the group dynamics that take place. Reporting should concentrate on what information has been utilized, what information is needed, and any difficulties in getting that information. The manager must be willing to assist the team with its assignments. Sometimes assistance may take the form of beginning a business research study to find out more on a particular issue. In other situations, it may mean getting appropriate clearance for information.

Basically, the manager needs to show continuing support for the team after it has started the project. The team has the responsibility to complete the project in a timely manner and to keep the manager informed on a routine basis. Good communication is the key to success.

Another question that managers ask is, "Who should be on the team?" Here there are two paths that can be taken. On a new product, the team should consist of development people, marketing people, business research people, QA, and manufacturing. This team enables the tracing of the project from concept to product launch. Each person has a key role. Development will bring the concept to reality and assess the feasibility; marketing will determine the market for the product based on customer needs; business research will determine what studies need to be done to address the unknown categories; QA will determine what techniques are available to ensure the quality and develop new techniques as appropriate; and manufacturing will assess the capability of the current equipment, to see if new tooling is required.

A product-improvement team will generally consist of development, marketing, QA, and manufacturing people. This team is smaller because the product is already available. Improvements are defined by customer needs, which are obtained from marketing. Development, QA, and manufacturing will assess the availability of current technologies, techniques, and equipment in order to determine the resources needed to bring about the improvements. Resources in this case are defined as money, time, people, and equipment. Of these, the limiting resource is usually time. That is the hardest resource to utilize effectively and is generally the one that is working most against any project.

As a team, this information will be assimilated into the various matrices so that decisions can be made and the trade-offs identified.

Generally, the first project selected should be rather limited in scope. An improvement project is best since it has a foundation of existing information available. It is somewhat easy to identify the team members and to define the scope of the project. There is usually some information on customer expectations and competing products. This gives the team an easy introduction to the concepts and techniques of QFD, and it also provides a quick completion of the initial matrix. This results in some feeling of accomplishment and makes further sessions flow better.

Ideally, prior to the QFD exercise, the team should meet to accomplish two things: first, to learn the scope of the project and second, to learn something about the QFD process. Understanding the scope and expectations helps establish priorities, so as the QFD process is explained, the team can get a feel for the amount of effort and information that will be necessary to complete the task. At the completion of this meeting, the team members will be told when the first session will begin and how long it will last. This helps establish the importance of the project. This initial meeting is also a good opportunity for the team to get together and meet each other. In many cases, there are "new" people on the team. Understanding who is there and where they work reinforces the team approach to the process.

Prior to the first team meeting, the sponsor and the facilitator need to establish the format, duration of the meeting, and the time frame for the various deliverables. This is important in order to establish checkpoints along the way to measure how effective the QFD process is. Failure to complete this step can result in divergent paths, which slow down the process as well as create an uncertainty as to when the project will be completed. The format found to be most effective is to first review what you want to accomplish in the meeting, resolve as best as possible any unanswered questions, and finally get into the process. This is only one way to set the environment. It is necessary for the facilitator to be flexible enough to vary the format as the situation demands.

The duration of the sessions is also critical to the success of the project. Workshops might last two days, one day (eight hours), four hours, or two hours. The key point to remember is, what are you trying to accomplish? If you are bringing in people from around the world, then a two-day workshop may be necessary. If the team is working on a product improvement and members have worked together prior to this, a two-hour session twice a week may accomplish just as much with less disruption to the work schedule.

Several things have to be taken into consideration.

- What are you trying to accomplish in the session? If you are trying to develop a complete matrix in a short period of time, a compressed schedule may be necessary. This means that some up-front planning needs to be done. All team participants need to be informed of the scope of the project and what needs to be accomplished. They need to be told of the types of information that are needed so that they can come prepared to the sessions.

- How much time should be allocated? This is always a difficult question to answer. If everyone is prepared and understands the scope of the project, it will take between sixteen and twenty hours to develop the first matrix. This estimate is dependent upon how prepared the team members are. When information is readily available, the matrix can be completed. If there is not a lot of work done prior to the sessions, this number can double very quickly.

- Where are the participants coming from? This question can mean two things: what groups/organizations are to be included in the exercise and, in a large company, what manufacturing sites and marketing regions need to be involved? The first question has been addressed earlier in this chapter. The second will dictate how the session will be put together. If there are people coming from other locations, then in order to minimize cost, there needs to be a workshop run over two days, so that the initial matrix can be completed with all members of the team present. If all members of the team are from the same location, then the time can be broken up into four- or two-hour blocks.

- Two-hour blocks tend to be easier to schedule and you can accomplish one segment of the matrix in that time. For example, at the end of the first two-hour meeting, the team will have completed the customer requirements side of the matrix. In the second session, the team will develop the manufacturing requirements. These would then be joined together to form the matrix. The third session will look at the planning matrix, and so on.

The advantage of two-hour sessions is that there is time to gather information between the meetings. This helps to ensure that the right type of information is utilized in building the matrix. This also enables the team to initiate studies that will fill in any gaps that have been uncovered. Another advantage is that it helps the team remain on focus and not be sidetracked into other discussions. It is easy to let the team wander off when it is perceived that there is "plenty of time" to get the segment that is being worked on finished.

The role of the facilitator is primarily to get the QFD exercise completed in the most efficient manner. This means that all team members know what is expected, all information is compiled quickly, consensus is reached in a timely manner, and cooperation inside and outside of the sessions take place. As stated in Chapter 1, the benefits of QFD are primarily communication, teamwork, and documentation. The facilitator is there to enhance all three. Frequently, the facilitator can speed up the documentation by having an apprentice facilitator at the session inputting the data into a portable computer. This enables sessions to be held more closely together, for example, two or three times a week, and still have hard-copy documentation available at each session.

The facilitator needs to maintain a close relationship with the team leader. The two need to meet prior to the QFD exercise to review the process, determine the scope of the project, recommend what groups/organizations should be on the team, determine the length of the sessions, and help prepare the initial letter explaining what is to be accomplished and what is needed for the exercise to be successful.

During the initial meeting, the facilitator needs to ensure that everyone understands the process while the team leader further defines the scope of the project.

After each session, the facilitator should review with the team leader the outcome of the meeting, review the strategy for the next meeting, and recommend methods to gather additional information based on the data received.

3

How to Obtain Customer Information

When managers are told that they need to define customer requirements, their first response tends to be, "Well, I'll get our Business Research people on it." What happens then is the manager finds out that the Business Research Organization already has a full slate of projects and that they will try to "fit him in." This usually results in the manager doing one of three things: he or she waits for customer information, goes to upper management to get Business Research's priorities changed, or goes out to get the data him- or herself. There are pros and cons to each of these approaches.

In this chapter, another approach is presented that has been found to promote better communication and information between different organizations in a business. The chart in Figure 4 summarizes the various types of customer information resident in most businesses. It may look confusing at first glance, but once understood, it provides a good road map.

Customer information comes from a variety of sources; some are solicited and some are not, some are quantitative or measurable and some are qualitative, some are obtained in a structured manner and some are obtained in a random manner.

Figure 4
TYPES OF CUSTOMER INFORMATION

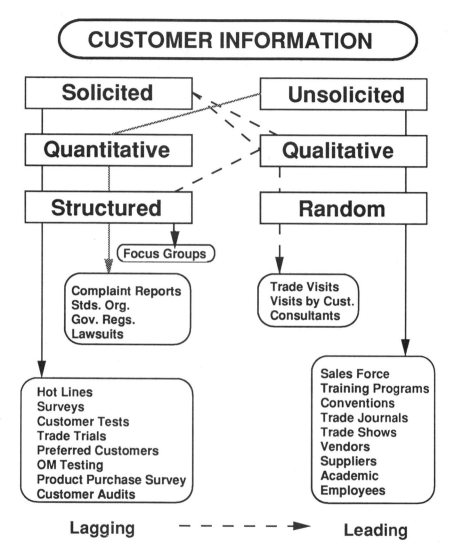

Another way of saying this is that information can come in in an active sense because the company is looking for it or, because it is told to (unsolicited), in measurable (quantitative) or subjective (qualitative) forms, on a routine (structured) or haphazard (random) manner.

Data that are solicited, quantitative, and structured tend to take the form of customer surveys, market surveys, trade trials, working with preferred customers, analyzing other manufacturers' products, and buying back products from the field. This information is valuable because it identifies where the company is currently positioned in the marketplace and shows strengths and weaknesses with generally good data. The drawback is that these data are lagging data because they tell how the company is currently doing, not where the company is going.

Unsolicited data that are quantitative and structured also have this weakness. These data tend to take the form of complaint reports or are received from various governmental and regulatory agencies as requirements or standards. They also can come in the form of lawsuits. This unsolicited information is not generally liked, but it is something that has to be conformed to. The important thing to remember is that it is information.

The last type of lagging information is solicited but is more subjective in nature and is obtained in a structured manner by means of focus groups. Focus groups are meetings with various industry leaders that are run by trained facilitators to find out likes, dislikes, trends, and opinions about current and future products. The meetings are usually videotaped behind a two-way mirror, and the tapes are reviewed by the sponsoring company. This information is valuable in that opinions and desires are captured in the customer's words, not translated or filtered by some technical person. The unfortunate thing is that this information is rarely seen by nonmarketing people. Consequently, most of this valuable data is "lost" to the people who can make the most of it—development and manufacturing.

There is other information available, which can be thought of as leading indicators of where the technology is going. It can be solicited by a company looking for qualitative information in a random manner in the form of trade visits to customers and noncustomers, by bringing customers into the manufacturing and development areas to talk about what they would like to see and by talking to the various consultants in the field.

Another source is the unsolicited information that comes in from salespeople, service reps, training programs, conventions and trade shows, and various trade journals, as well as from current suppliers, academic programs, and what employers hear from their friends and neighbors. All these unsolicited data are tapped into when utilizing conventional methods.

QFD can provide an access to this information and place it in a structure that can utilize it. Then the team has the foundation to see not only what the customers say that they want, but what the customers want but are not expressing. These are the things that lead to what has been called "exciting quality" and "customer delight." Bob King of GOAL/QPC talks about this in his paper "Listening to the Voice of the Customer: Using the Quality Function Deployment System."[1]

When trying to determine customer requirements, each QFD team is challenged to look beyond what the customer is asking for in order to determine the "expected quality" and the "exciting quality" demands. Traditionally, "expected quality" is not expressed because people are only upset when these characteristics are not available. Some call this type of quality the "price of admission." If you are going to do business, the "expected quality" has to be there or no one will buy the product.

"Exciting quality" is where innovation comes into QFD. These quality characteristics are the improvements that are not asked for but which create "delight" in the customer. An example is the cup holder in cars. Ten years ago, few American cars had

---

[1]King, Bob. "Listening to the Voice of the Customer: Using the Quality Function Deployment System." *National Productivity Review,* Summer 1987, pp. 227-281.

cup holders, while many Japanese cars did. Customers did not look for this characteristic in a car, but were pleased when they got it. This resulted in customers looking for cars with cup holders when they were shopping around for a new car. The "delight" or "exciting" characteristic now became a demand. The cycle begins again, looking to find the new "exciting" quality characteristic. This is one way to ensure that the company is looking for continuous improvement.

Stuart Pugh developed a technique to look at new technology concepts. He utilizes a matrix to check how each new technology meets the expected quality characteristics. The team utilizes the current product as a benchmark to determine if the new technology is better or worse. If better, it is assessed for technical difficulty to see how feasible it is for implementation. This is another area for creating "customer delight." New technological concepts could create additional benefits, which can win over customers.

In summary, various ways to get customer information have been presented. Some is information that is lagging and some is looking forward. All information must be looked at for what it tells you and what it omits. The team must seek out the "expected" quality characteristics and the "exciting" quality characteristics. Only by doing so will new and innovative approaches be found and verified in a QFD matrix.

4

Putting It All Together

You have management support, you have put together a team, and you are now ready to begin the first meeting. Right? Wrong! As you begin the first session, two things need to be done to set the tone. First, the team leader should define the scope of the project, and second, you should spend some time defining who is the customer. In many cases, there is more than one customer. The obvious customer is the ultimate consumer, but there may be some intermediate customers such as field service organizations that install and maintain machines, printers who buy inks and other consumables, and so on.

For example, if you were making a copier, your ultimate customer would be the person who receives a copy of some document. The intermediate customers are the organization that maintains the equipment, the key operator who runs the machine, and the person who requests to have copies made. Each customer has a different set of needs that are important to him or herself. The team should look at each of those customers.

Once the customer issue has been decided, then starts the brainstorming session to determine what the customer requirements are. The usual rules of brainstorming are used; for example, all ideas are valid. Let the Affinity Diagram tool do the sorting. The Affinity Diagram is a tool that enables many ideas to be organized into categories based on national relationships. (See Part 2 of the book for more information.)

The facilitator should be recording the ideas. Perhaps a second facilitator can write the ideas on 3x5 cards while the facilitator is writing on a pad. This not only saves time, because the facilitator does not have to transcribe onto the cards, but it also provides a learning experience for someone who has just learned how to do QFD. A co-facilitator and a facilitator can switch roles during sessions so that the team always has a "fresh" facilitator.

The most common mistake new facilitators make in this process is talking too much. In brainstorming, there are times when silence takes over. It is important to recognize that people need this time to formulate ideas. New facilitators tend to offer alternative approaches to "stimulate" ideas. There are times when this is necessary, but only when a period of prolonged silence has elapsed, between two and three minutes. Any sooner tends to break thought patterns that were being formulated and may be lost for the exercise. The facilitator has to have patience, since some of these team members have never worked together before. The facilitator is there to help, not distract.

Another task that the facilitator has is to seek clarification on terms, keeping the team focused on what the customer would ask for, not the technical translation. Let the matrix do all the translation! When not sure of a term, ask, "How would the customer say that?" This is a simple way to keep the team focused on the customer's perspective. In certain professions, the terms may be similar, so it helps if the facilitator can learn something about the product and uses prior to the initial session.

One thing that is helpful is to write on the back of the 3x5 cards some words that summarize some of the discussion so that when the team asks at a future time, "What did we mean by that?", you can briefly explain the discussion that took place. This is extremely beneficial when team meetings are stretched out over weeks.

When the brainstorming is completed, the group is briefed on how to do the Affinity sorting. Explain that for the first five minutes, there will be no talking. Each person will place the cards that seem to be related into groups. Then send the team on a break and scatter the cards on a large table. The key word is "scatter" since the team will be around the table doing the sorting. Get the team started and watch what happens. In a short time, a majority of cards will be grouped. Then ask the team to find a card or word that best describes each group. This becomes the overall description for each grouping.

By using a different color marker for the headings, you can easily differentiate groupings when working after the session constructing the Tree Diagram. Another way is to number the cards by grouping. The only difficulty in doing this is keeping large groups and sub-groups in order. As the grouping is identified, lay the cards on the table in the same order that the team has grouped them. This creates a visual image of the Tree Diagram as it will look on the matrix.

The next step is to develop the technical characteristics part of the matrix. If you are starting on a simple project like a product improvement, the easiest way to start is by listing all the technical characteristics that are measured or evaluated in the current product. By starting this way, the matrix can be used to see how well the current quality system assures that the product meets customer needs. In many cases, this is the first time anyone has documented all the tests that are performed on the manufactured product. This will amaze some team members.

After everything is listed, the team will go through another card sorting based on the technical characteristics. This will be an interesting exercise, since some tests generally are not grouped where some technical people expect them to be. This tends to create discussion on what should be looked at and usually results in more cards being made up to "fill in the gaps."

The next step is sometimes considered a "reality check." A planning matrix is now developed, which contains the importance rating, where the current product stands "from the customer's perspective," where other manufacturers stand, where you want to position yourself with the improved product, a scale-up factor that is calculated from the scale-up and current position, a sales point, and a weight.

The importance rating can use any scale, but is generally a 1 (low) to 5 (high) scale. There are some people who state that "everything is important." In the purest sense, that is true, but there is also a hierarchy of importance. A good example of this is child-proof bottle caps for medications. Safety is the most important requirement, followed by clear instructions, and ease of opening. Small children cannot open these caps, so the primary importance was achieved, but not all of them are easy to open, so trade-offs were made that prevented this requirement from being fulfilled.

The current product standing also tends to be rated on a scale of 1 (low) to 5 (high), although any scale can be used. This rating is what the customer thinks of your product or of the product that you are replacing if you are looking at a new product. It is important to emphasize to the team that "this is how the customer feels." Benchmarking comes later. Focus groups and customer complaints are the most common sources for this information.

The next columns look at the competition. They should be identified so that everyone on the team knows who the competition is. It is imperative that the same scale be used here as was used for your product. This is where some insight can be gained—how the competition sees weaknesses in your product. It is also the start for you to see how you can develop some launch strategies to emphasize strengths and improvements in your product.

The target column is on the same scale as those for you and your competition. The decision here is to improve, remain equal to the competition, or remain behind the competition. Improvement is desired in most companies, but may not be attainable if the competition is considered best. In these cases, parity may be the only option, unless due to some constraints you have to take a lesser position. Reality sometimes forces this decision, so the team must be made aware of constraints to the design if there are any. In one circumstance, a project was canceled after this stage because the design restrictions revealed that the new product was inferior in a critical element to that of the competition.

The scale-up is the ratio of the target to the current product. This is an indicator of how much effort is needed to meet the target. The higher the number, the more effort is needed. This number must be looked at carefully. Where you start out on a product maturity curve can result in a low scale-up where it should really be a high number. An example helps explain this point. A product was rated 4 on the current product evaluation; the target was 5. The scale-up was 1.25 (5/4). The product was considered very mature, so to move to the target was a major task, which would tie up many resources. This was contracted by a packaging change where the current product was rated 2 and the target was to match the competition at level 4. The scale-up here was 2.0 (4/2). The packaging change was relatively easy to do at minimum expense. The scale-up would lead one to believe that more resources were needed for the packaging change as opposed to the other characteristic where just the opposite was true! Look at the level from which you are starting to determine the validity of the scale-up. If necessary, re-evaluate your ratings to see if they are realistic.

The sales point is a measure of how "sellable" a particular requirement is. If you are considered the "best," it is a high sales point and should be included in your sales literature and training. If it is not "best," it should be a low sales point. Promote your best characteristics. Use the others to help the sale, but emphasize the "best." This column is used to develop marketing strategies.

The weight is a calculation that takes importance, scale-up, and sales point. From this, the weights can be ranked on a Pareto chart from high to low. This can then be used as a road map for development activities. The resources can be allocated to the list in order of priorities.

The next step is filling in the middle area of the matrix. In this step, there tends to be a great deal of discussion. Now the team will identify the relationships between the customer requirements and the technical requirements, which can take a lot of time if you have a large matrix. The relationships are defined as strong, medium/some, and weak/possible relationships. Symbols are usually used in order to aid in the recognition of patterns. Numbers are substituted in at a later time to calculate weight at the bottom of the matrix. This is where QFD provides a translation of what the customer requirements mean to the manufacturer. This shows the synergistic effect meeting the customer requirements; by meeting one requirement, other requirements can also be satisfied.

Now the bottom of the matrix can be completed. Any benchmarking that has been done can be added. This is where your company is compared to the competition using all the technical characteristics. Now the team can look for inconsistencies between what the customers perceive and what has been measured. This can show areas that need improvement, where marketing strategies need improvement, and verification of being "best." On all characteristics, some assessment of technical difficulty is now needed. This is usually done on a 1 (low) to 5 (high) scale, which will help in the determination of the development efforts.

The last two rows will be two weights; the first will be the row weight and the second a scaled weight. The row weight is a simple total of the relationships in each column. A strong relation equals 9, a medium equals 3, and a weak equals 1. This total shows the impact of the technical characteristics on the customer requirements.

The second weight uses the planning weight from the planning matrix with the relationships. Each relationship is multiplied by the planning weight and then totaled. This scaled weight is then totaled and put into a Pareto chart to show which technical characteristics are most important in meeting the customer requirements. This, in addition to the technical difficulty rating, helps determine where to allocate resources.

The last thing to do is to look at the peak of the "house." This is where the technical characteristics are evaluated against themselves. Here you are looking for positive and negative relationships. This is where the trade-offs are identified. This is valuable because in most cases, these have not been documented prior to this time.

At this point you have completed the foundation for all future matrices. In some cases, your QFD exercise will stop because of some business decision (for example, a decision to cancel the project). In many cases, you will continue to other matrices. Once the first one is completed, you and the team will have mastered the mechanics of QFD, and the rest of the matrices will develop more quickly. They will follow the same sequence, usually with a more abbreviated set of steps since you are now narrowing your focus to the critical areas for work.

5

Tools to Complete the Task

When a QFD exercise is started, the facilitator has a myriad of tools available to complete the task. The most commonly used tools are known as the Seven New Planning Tools, but there are others such as Value Analysis, Experimental Design, SPC tools, and so on, which can also be used. This chapter will look at a number of tools to show how they can be applied in the course of a QFD study.

The Seven New Planning Tools are a set of tools used for initial planning and QFD matrix generation (Affinity Diagram, Tree Diagram, and Matrix Diagrams); relationship understanding (Interrelationship Digraph and Matrix Data Analysis); and decision criteria (Process Decision Program Chart and Arrow Diagrams). These tools are explained in detail in Part 2 of this book.

The Affinity Diagram, also known as the KJ Method, is used to unlock creative thinking. This method looks for a different way of describing some relationship. When done correctly, it discourages the "pigeonholing" that some technical people tend to do, and it encourages team members to group items into similar sets, while not challenging the choice. Each person can regroup items as he or she sees fit. Generally, the team might spend the first five minutes grouping in silence. When the initial flurry of activity has slowed down, the talking takes place, usually to clarify the odd cards.

The Tree Diagram is used to define the hierarchy of tasks needed to be completed. It also defines the broader description of what is going on. You strive to go down to a level of detail that helps everyone understand what is going on. When that is done, go back up to see if there is agreement on the higher-level description. This check is important in case the decision is made to use the higher levels for the relationships. In some respects, the Tree Diagram is similar to a Cause-and-Effect Diagram in that it helps funnel the groupings arrived at in the Affinity Diagram into a logical sequence. It also identifies possible gaps when looking at the sequence when it does not logically flow from one step to another.

Matrix Diagrams are simply the joining of two sets of Tree Diagrams. This is the simplest of matrices and the most common to the QFD process. This forms the "body" of the QFD chart, which looks at the various relationships. By design, it forces the team to consider all aspects of the Tree Diagrams with each other, which then sheds some new perspectives on how things are looked at.

The Interrelationship Digraph takes some issue and develops a flow of logical steps. This is used to show the logical progression of steps needed to make something happen. This can be helpful in understanding how things relate, but experience has shown limited usefulness in a QFD study. The time required to develop it may be more productive when spent working on other matrices.

Matrix Data Analysis is a complex data analysis set of techniques that looks at many relationships and responses simultaneously. This is also known as Multivariate Statistics or Multivariate Analysis. Principle Component Analysis is the most popular technique, but others such as Factor Analysis, Discriminant Analysis, Cluster Analysis, Multidimensional Scaling, and Multidimensional Contingency Tables can also be used. These techniques assume that the responses are not independent of the predictor variables. They look to define the relationships in a truer description of what really happens. These are not trivial calculations, and one should exercise caution when interpreting the results. Computer packages like SAS have made these techniques more available to the nonstatistician, but someone familiar with these techniques should help when using them. While helpful in describing how things happen, explaining these techniques to nonstatistically oriented people only creates confusion. Care should be exercised in the utilization of these techniques.

Process Decision Program Charts (PDPC) attempt to look at every possible outcome and decision that can occur. This chart can be considered the opposite of a Failure Mode and Effect Analysis (FMEA), because it starts at the broad perspective and narrows the possible outcomes. This chart can become quite large very quickly. A good approach is to focus on an area that is considered critical and develop that area first.

Arrow Diagrams are variations of PERT charts. They help define how the subtasks interact in a new product development cycle. The key is knowing all the tasks that are necessary. In a generic sense this may be possible, but many times there are things we just do not know about a new product.

Stuart Pugh developed an innovative technique called concept selection, which has been adapted in many QFD studies. This concept started with the premise that the selection of the "best" concept is more difficult than the selection of the wrong concept, due to a state known as "conceptual weakness." Conceptual weakness occurs in two ways: (1) a weak design in general and (2) a strong design, but one that is not well thought out so it is subject to debate, which leads to lesser designs being chosen.

Pugh's approach is to compare all possible solutions with the same level of detail in a matrix format. The top of the matrix has the concepts, and the side has the criteria all the concepts will be evaluated against. Included in the top is the current standard as well as sketches of each concept. The scale used is simple: + means better than current standard, − means less than current standard, and s or = means the same as current standard. Total all scales (+, −, and =). If more than one concept is superior, then look at the superior concepts with the + rows removed. Do this until one concept emerges. If one concept does not emerge, then change the standard and re-evaluate the concepts. A check is to take the strong concept and resume the matrix to see if the results change. This accomplishes a number of things: better insight into the specifications, a better understanding of the problem, better ideas of alternative solutions, the identification of potential interactions, and an understanding of why one concept is better than another. This pattern can be repeated as long as necessary to obtain the level of detail necessary to develop specification.

Value Analysis (or Value Engineering) is another tool that has been effectively utilized. Traditionally, it has been used to identify the function of a product that adds value and to provide products at the lowest total cost. Value is determined by looking at both the positive and negative aspects of each function. A typical approach is to have each team understand the purpose of each component of a product. Then each function is evaluated by a cost-benefit factor, which is plotted on a graph that has importance on the y-axis and cost on the x-axis. A 45-degree line is drawn, and items below that line are targeted for improvement. These targets can then use Pugh's new concept selection for determining an alternative design. The new concepts are then evaluated for technical feasibility, cost, reliability, safety, environmental concerns, and quality. The "best value" design is then chosen to be implemented.

Traditional experimental designs and Taguchi techniques are tools utilized to understand the relationships in the body of the QFD matrix. There is much controversy among statisticians as to which tools do the job best. There are situations where both philosophies can be effectively utilized. Use whatever works best for what you are trying to understand. Through this understanding will come more efficient manufacturing operations, which will result in better utilization of the many SPC techniques so that higher-quality products will be produced at minimum cost.

All the tools discussed so far have been somewhat technical in nature. There are some behavioral techniques that the QFD facilitator has available to use. The first tool is one of feedback to the team. In many cases, the project team has never worked together as a group. Rather, members may have interacted individually to contribute some aspect. Group dynamics can be powerful in generating ideas that were not considered before. These ideas may be good and may not be so good. Feedback needs to be given so that each team member feels that he or she has positively contributed to the process. Failure to do so will tend to cause a team member to either stop coming or to show up and not contribute. Feedback can be given during the group session or individually after the session. It is important to see how each person feels the process is working. QFD is a powerful team-building tool when all team members contribute to the process.

Another tool is observation. See how others react to comments, and do not let one person dominate. In many technical groups, there tend to be some strong individuals. It is critical to get all team members to contribute.

The next tool is listening. It is important to capture the thought behind each requirement as it is discussed. Try to capture this through "key words," so that when the team meets again and asks, "Why is this here?" you can summarize the discussion and keep the process moving. This is critical when there is a group of meetings spread out over a period of time.

QFD exercises generate a wealth of information. The biggest obstacle in documenting each exercise is the time spent making up matrices. A number of companies have looked at developing computer software to make this task easier. Some companies have used various CAD systems to generate the matrices. These can be very impressive charts but may not be easy to modify when a change is required. Some companies use spread sheets like LOTUS & DB III to generate charts. These have better flexibility with respect to modifications, but can run into problems with space on the computer. Some companies have developed software for their internal use and for their suppliers, which do a pretty good job but are only available to a limited group of people. Some third-party software companies are also starting to develop software like QFD CAPTURE.

As software becomes available, QFD will become more ingrained in American industry. The critical element for each company is to evaluate how QFD is utilized. Some packages do not allow for modification of formulas or requirements. If you have an alternative way of scaling, then you need that capability. One point to remember is that the Japanese have not developed any commercially available software because they found that each QFD exercise is unique. Each application is just different enough that a structure that some software could provide could also inhibit some creativity. You may find that a framework such as LOTUS is helpful in providing flexible applications but that no two matrices look the same. It is at this time a give-and-take situation. Find what works best for your environment and use whatever tools are available.

6

Variations on a Theme

People say that QFD needs to be tailored to the project. What do they really mean? In this chapter, various alternatives to the "traditional" methods will be examined, which will show how flexibly QFD can be applied.

The first area to be looked at is the area of technical characteristics. What can be put in the section? Traditionally, we look at things that are measured so that specifications can be developed. Other applications require different characteristics; for example, one project looked at how the quality of work was affected by the manufacturing process. The customer requirements were things like safety, communications, materials, and so on. The top of the matrix looked at the manufacturing process steps. This was an interesting exercise in that when completed, the division reorganized to improve communication and product flow. Some management jobs were consolidated, and manufacturing costs were reduced by elimination of redundant testing.

In a project-planning QFD, the top of the matrix listed all major development projects, which led to better coordination of activities so that projects did not overlap. A variation of this looked at the customer's process steps to see how products could be developed to reduce the customer's work.

A staff group listed the services that they provided their customers so that the staff group could measure the effectiveness of its work. This resulted in improved communications between the staff group and clients.

In the planning matrix, there are a number of variations that can be used. Traditionally, the importance rating is scaled from 1 (low) to 5 (high). On some projects, this scale is changed to 1 (low) to 10 (high) when a finer breakdown is needed. One team went from a -1 (negative impact) to a 5 (very high value) to rate customer importance. Some teams place words around the numbers, such as $1-$not important, $2-$somewhat important, $3-$important, $4-$very important, and $5-$if not there, I will not buy the product. The important thing to do is to determine the scale before starting the exercise.

When looking at technical characteristics and new concepts, traditionally a $+$, $-$, or $=$ scale is used. There are times when the relationship is not known and a ? is used. This is done so that designed experiments can be set up on concepts that show potential. In general, ? can be used to highlight areas that need more information. Caution needs to be exercised in doing this so that the teams do not abuse this option.

QFD studies have been used to justify new equipment by using the matrix to show how the equipment can reduce bottlenecks, eliminate tasks, and improve quality.

It is often difficult to explain uses without a case study. There are not many case studies on QFD, because many of the companies that are using the technique find that it gives them a competitive advantage. Most studies go only so far and then are stopped. This fact in itself is an indicator of the value of QFD. When does a QFD exercise end? Purists may say when a product is launched into the marketplace. That is so in some cases, but there are also times when a QFD exercise can be ended early. A strategic planning QFD may just focus on the first matrix. Completion of that can predict customer trends and what potential impact a new product can have in the marketplace. It can also address current technical capabilities so that you have some idea of the skills your company should be looking for.

A new product development QFD can be ended when the manufacturing feasibility is determined. The ability to make a product is a critical step. The QFD may have identified the major restrictions for making the product. To move beyond that may need a re-direction in the overall business plan.

Staff and service QFDs will end when they complete the scope of the project. This may be better communication, the establishment of routine customer feedback, or the development of quality measure. One team looked at better ways to obtain customer information. The QFD produced some better approaches.

Regardless of what the scope is, the facilitator and project leader need to determine when a project ends. This is then presented to the team and the team begins work on a report, which will show the conclusions reached by the team with recommendations for the next step. This report will be presented by the team to the appropriate management groups. This is also an opportunity for the team to let management know what they think about the QFD process. This is important and is usually overlooked by most teams. Management needs to know whether or not QFD is a viable process, regardless of the outcome. The project may be canceled because of the information that the QFD process revealed. If the process helped in that decision, management needs to know that!

In addition, it is also recommended that facilitators meet with the team and the project leader to evaluate the process. This is necessary if you expect the QFD process to be continually improving. Lessons learned can save time and money later and can also create the opportunity for creative innovations to the QFD methodology. It is important to remember that the facilitator's customers are the project leader and the team.

One use of QFD that is still in the early stages is as a Supplier Partnership tool. When trying to establish Supplier Partnerships, there are some phases that customers and suppliers go through. This can be similar to an interpersonal relationship resulting in marriage. The first meeting is like a "date." Both parties try to put their best foot forward. When an initial contract is made, this is the "going steady" phase. Both parties are making a limited commitment to see if the initial impressions are true. "Engagement" occurs when longer-term contracts are made based on past performance. "Marriage" happens with a long-term contract and the invitation to assist in new product development. QFD can begin at any one of the phases since you are your supplier's customer. It does require some investment in terms of time and people, but in the long run you will benefit with better product when you want it.

7

QFD as a System

When people talk about QFD, the discussion is generally about matrices. One rarely hears about QFD as a systems approach to product planning. When a systems approach is used, management acceptance is obtained because managers can see the value. Figure 5 shows a generic flow of the QFD system model. The value of this model is that it shows the integration of the various quality tools with a set of deliverables using QFD as the framework.

The cornerstone of the system is the Customer Requirements and Engineering/Technical Features matrix. From this all other matrices and studies are generated. From this the next matrix shows the Engineering Features and the Applied Technologies. This is where new concepts are evaluated. These two matrices result in a Customer Needs Document, a Concept Document, an assessment of the technical requirements, and the identification of trade-offs. The Quality Tools utilized in these matrices (and in all the matrices) are Affinity Diagrams, Brainstorming, Tree Diagrams, Matrix Development, and Pareto Charts.

Once the concept is approved, the next matrix looks at the Applied Technologies and the Manufacturing Steps. Here is where Quality Tools like Cause and Effect Diagrams, FMEA, PDPC, Experimental Designs, Taguchi Experiments, and Interrelation-

Figure 5
QFD SYSTEMS FLOW MODEL

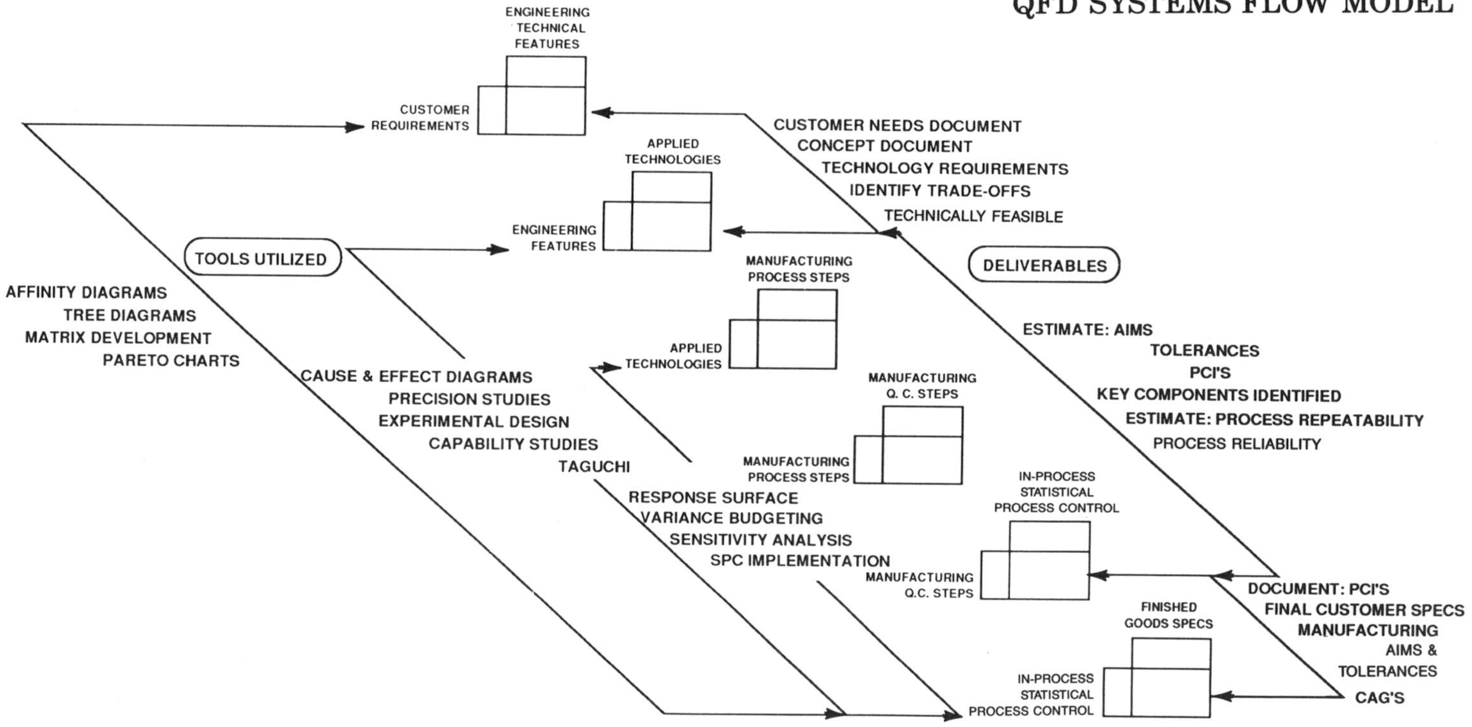

DELIVERABLES

CUSTOMER NEEDS DOCUMENT
CONCEPT DOCUMENT
TECHNOLOGY REQUIREMENTS
IDENTIFY TRADE-OFFS
TECHNICALLY FEASIBLE

ESTIMATE: AIMS
TOLERANCES
PCI'S
KEY COMPONENTS IDENTIFIED
ESTIMATE: PROCESS REPEATABILITY
PROCESS RELIABILITY

DOCUMENT: PCI'S
FINAL CUSTOMER SPECS
MANUFACTURING
AIMS &
TOLERANCES
CAG'S

FINISHED
GOODS SPECS

IN-PROCESS
STATISTICAL
PROCESS CONTROL

IN-PROCESS
STATISTICAL
PROCESS CONTROL

MANUFACTURING
Q.C. STEPS

MANUFACTURING
Q. C. STEPS

MANUFACTURING
PROCESS STEPS

MANUFACTURING
PROCESS STEPS

APPLIED
TECHNOLOGIES

APPLIED
TECHNOLOGIES

ENGINEERING
TECHNICAL
FEATURES

ENGINEERING
FEATURES

CUSTOMER
REQUIREMENTS

RESPONSE SURFACE
VARIANCE BUDGETING
SENSITIVITY ANALYSIS
SPC IMPLEMENTATION

CAUSE & EFFECT DIAGRAMS
PRECISION STUDIES
EXPERIMENTAL DESIGN
CAPABILITY STUDIES
TAGUCHI

TOOLS UTILIZED

AFFINITY DIAGRAMS
TREE DIAGRAMS
MATRIX DEVELOPMENT
PARETO CHARTS

40

ship Digraphs come into play. At this point you are looking to estimate aims and tolerances and PCIs. You want to identify key variables in the manufacturing process.

The next matrix looks at the Manufacturing Process Steps and the Manufacturing QC Steps. Now you start your SPC implementation and run capability studies, as well as continue any experimentation started earlier. Estimates are being obtained for process repeatability and reliability as well as to determine your testing capability. This is where process optimization starts taking place.

The next matrix looks at the Manufacturing QC Steps and the Process SPC parameters. This can be considered a check to ensure that the right process variables are being used. Any SPC plan that does not check itself periodically is potentially destined for trouble. This matrix (as with all the others) can have the Customer Requirements used in place of the Manufacturing QC Steps. In this way, you can directly see the relationship of the SPC variables to what the customer wants.

The final matrix in the flow looks at the Process SPC parameters and the Final Customer Specifications. At this point, PCIs are documented, the SPC implementation is in place, specifications are confirmed, manufacturing aims and tolerances are established, and Corrective Action Guidelines are in place. At the same time, product-launch strategies are finalized and the company prepares for the release of the product.

Using this flow model, managers can see the potential strengths in utilizing QFD. As the project progresses, other matrices may be utilized to better clarify requirements. The advantage of the model is to show how QFD flows from design concepts to a manufactured product. This process aids the difficult transition of bringing a product from development to manufacturing. It also brings all the necessary information to manufacturing so that the line operator is capable of running the process as necessary to produce the highest quality product.

PART TWO

Seven New Planning Tools for Quality and Productivity Improvement

INTRODUCTION

In 1950, Dr. W. Edwards Deming drew the following diagram on the blackboard during his first meeting with the Japanese Union of Scientists and Engineers (J.U.S.E.):

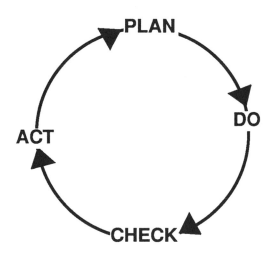

Dr. Deming called it the Shewhart Cycle. It has since become known as the Deming Cycle (or Plan-Do-Check-Act Cycle). It graphically describes the action steps we use every day to manage our lives and our businesses.

1. We **PLAN** what we want to accomplish over a period of time and what we are going to do to get there.

2. We **DO** something that furthers the goals and strategies developed in number 1.

3. We **CHECK** the results of our actions to make sure that there is a close fit between what we hoped to accomplish and what was actually achieved.

4. We **ACT** by making the changes that are needed in order to more closely achieve the initial goals or develop procedures to assure continuance of those plans that were successful.

Problems Implementing this Cycle.   Even though this is a *natural* process, its effective application in many U.S. companies has been limited by several major factors.

a. Since the days of Frederick Taylor the planning and evaluation functions have been separated from the *doers*. This was based on the principle that technical specialists (e.g., industrial engineers, Q A engineers) knew best how to plan to do a job while an unskilled workforce knew only how to execute the plan. This eventually led to the strict departmentalization of job functions that is so prevalent today in the U.S.

b. Planning has often been relegated to the "seat of the pants" approach. It has often been seen as either too theoretical to be of practical use or too detailed and mundane. It has not been viewed as being where the action is. There is the common perception "that doers are recognized and rewarded while planners just plan."

c. There has been a lack of available tools that make the job of planning simple and timely.

Impact of the Quality Revolution.   The Quality Revolution that is underway today is addressing these issues. The Deming Philosophy and the concept Total Quality Control (TQC) (or Company-Wide Quality Control [CWQC]) focus heavily on breaking down these organizational barriers to improvement. Based on these approaches Taylorism in many U.S. companies is an endangered species.

The "Seven New Planning Tools For Quality and Productivity Improvement" (hereafter The New Tools) finally provide every manager with the tools needed to make planning an effective and satisfying process. They also break down Taylor-type barriers by giving more individuals the ability to contribute to the planning step.

History of the New Tools

Most of the New Tools are not new at all. Rather, most of them have their roots in post World War II Operations Research work. Over the last fifteen years or so, the Japanese have taken these techniques from Operations Research and combined them with other tools to form a powerful planning cycle. The Japanese often take simple components and make the *whole* more powerful than any one of its *parts.* The most advanced Japanese companies are now taking advantage of this cycle to cut their design time drastically, plan for new product introductions, and plan their transition to CWQC. Since the New Tools were introduced in the U.S. through GOAL in 1984, a number of U.S. corporations have begun to use them with excellent results in a wide variety of applications.

What Are these Mysterious New Tools?

The New Tools are neither complex nor mysterious and include:

1. Affinity Diagram/KJ Method

2. Interrelationship Digraph

3. Tree Diagram/Systems Flow Diagram

4. Matrix Diagram

5. Matrix Data Analysis

6. Process Decision Program Chart (PDPC)

7. Arrow Diagram

Tool Description

The seven New Tools are as follows:

Affinity Diagram (KJ Method)–This tool gathers large amounts of language data (ideas, opinions, issues, etc.) and organizes it into groupings based on the natural relationship between each item. It is largely a creative rather than a logical process.

Interrelationship Digraph–This tool takes complex, multi-variable problems on desired outcomes and explores and displays all of the interrelated factors involved. It shows graphically the logical (and often causal) relationships between factors.

Tree Diagram/Systems Flow Diagram–This tool, which resembles a horizontal organization chart, systematically maps out the full range of tasks/methods needed to achieve every GOAL/purpose. The very structured process translates the most general goal into the practical implementation steps that need to occur.

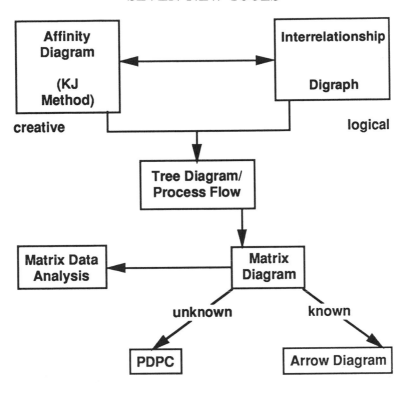

FIGURE 6

SEVEN NEW TOOLS

Matrix Diagram—This tool takes the necessary tasks (often from the Tree Diagram) and graphically displays their relationship with people/functions or other tasks. This is frequently used to determine who has responsibility for the different parts of an implementation plan.

Matrix Data Analysis—This is the most statistically sophisticated of the New Tools. Its graph shows the strength of the relationship between variables which have been statistically determined. This is frequently used in marketing and product research.

Process Decision Program Chart (PDPC)—This tool maps out every conceivable event and contingency that can occur when moving from a problem statement to the possible solutions. This is used to plan each possible chain of events that need to happen when the problem or goal is an unfamiliar one.

Arrow Diagram—This tool is used to plan the most appropriate schedule for any task and to control it effectively during its progress. This is closely related to the CPM and PERT Diagram methods. This is used when the task at hand is a familiar one with subtasks that are of a known duration.

Who Uses the Tools?   These New Tools have proven useful to virtually any level manager in a company. However, they seem particularly helpful for middle to upper managers. They seem to be particularly acceptable at these levels because they fill a void by the "Seven Old Tools." These old tools appear to have several disadvantages in the eyes of middle to upper-level managers.

1.   The "nonstatistical" charting techniques such as the Flow Chart, Check Sheets, Trend Charts, Pareto Charts and Cause and Effect Diagrams appear to be too simple. Therefore, someone *lower* should take care of such mundane activity.

2.   The more statistical tools such as Histograms, Scatter Diagrams, Control Charts (and eventually Designed Experiments) are too technical and should be done by specialists.

3.   The old tools are useful primarily for numerical data. Managers are often faced with verbal data. They often cannot effectively analyze or organize thoughts and words.

If these views are held, there is nothing left that is appropriate for these managers to use. However, the New Tools seem to be sophisticated enough to warrant their attention yet simple enough to be mastered in a fairly short time. Most importantly, the New Tools actually make the job of management under-standable and manageable.

The key to learning these tools lies in the following story. A young man carrying a violin case stopped a cab driver in New York City and asked him, "How do I get to Carnegie Hall?" The cab driver answered, "Kid, practice, practice, practice!"

AFFINITY DIAGRAM/KJ METHOD

Definition.   *This tool gathers large amounts of language data (ideas, opinions, issues, etc.) and organizes them into groupings based on the natural relationship between each item. It is largely a creative rather than a logical process.*

The biggest obstacle to planning for improvement is past success or failure. It is assumed that what worked or failed in the past will continue to do so in the future. We therefore perpetuate patterns of thinking that may or may not be appropriate. Continuous improvement requires that new logical patterns be explored at all times.

The KJ Method is an excellent way to get a group of people to react from the creative "gut level" rather than from the intellectual, logical level. It also takes these creative new thought patterns and efficiently organizes them for further elaboration. These thought patterns have seen teams produce and organize more than 100 ideas or issues in thirty to forty-five minutes. Think of how long that task would take using a traditional discussion process. It is not only efficient, however. It encourages *true* participation because every person's ideas find their way into the process. This differs from many discussions in which ideas are lost in the shuffle and are therefore never considered.

When to Use the Affinity Diagram/KJ Method

We have yet to find an issue for which KJ has not proven helpful. However, there are applications that are more natural than others. The "cleanest" use of KJ is in situations in which:

a. Facts or thoughts are in chaos. When issues seem too large or complex to grasp, try KJ to "map the geography" of the issue.

b. Breakthrough in traditional concepts is needed. When the only solutions are old solutions, try KJ to expand the team's thinking.

c. Support for a solution is essential for successful implementation.

KJ is **not** suggested for use when a problem: 1) is simple or 2) requires a very quick solution.

Construction of an Affinity Diagram/KJ Method

The most effective group to assemble to do a KJ is one that has the needed knowledge to uncover the various dimensions of the issue. It also seems to work most smoothly when the team is used to working together. This enables team members to speak in a type of shorthand because of their common experiences. There should be a maximum of six to eight members in the team.

The following are the most commonly used construction steps:

1. Phrase the issue to be considered. It works best when vaguely stated. For example, "What are the issues surrounding getting top management support for a CWQC process?" There should be no more explanation than that since more details may prejudice the responses in the "old direction."

2. The responses can be recorded in one of two ways:

    a. Recorded on a flip-chart pad and then transcribed onto small cards (e.g., 1″ x 3″), one idea per card.

    b. Recorded directly onto individual cards by a recorder or by the contributor him- or herself.

    Note: It must be stressed that ideas should be concise and recorded exactly as stated. The aim should be to capture the essence of the thought.

Figure 7
KJ/AFFINITY DIAGRAM
Ideal Child's Car Safety System

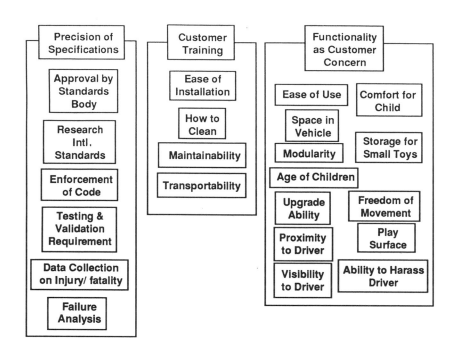

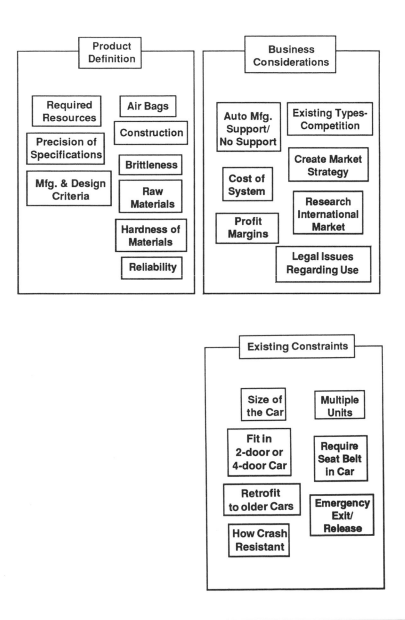

Product Definition

Required Resources	Air Bags
	Construction
Precision of Specifications	
	Brittleness
Mfg. & Design Criteria	Raw Materials
	Hardness of Materials
	Reliability

Business Considerations

Auto Mfg. Support/ No Support	Existing Types-Competition
	Create Market Strategy
Cost of System	
	Research International Market
Profit Margins	
	Legal Issues Regarding Use

Existing Constraints

Size of the Car	Multiple Units
Fit in 2-door or 4-door Car	Require Seat Belt in Car
Retrofit to older Cars	Emergency Exit/ Release
How Crash Resistant	

Figure 8
KJ/AFFINITY DIAGRAM
Ideal Child's Car Safety System

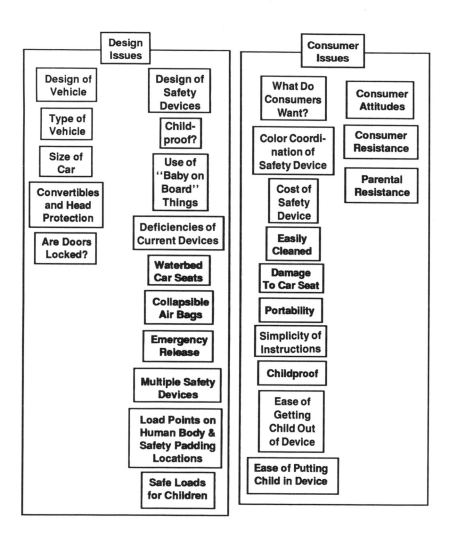

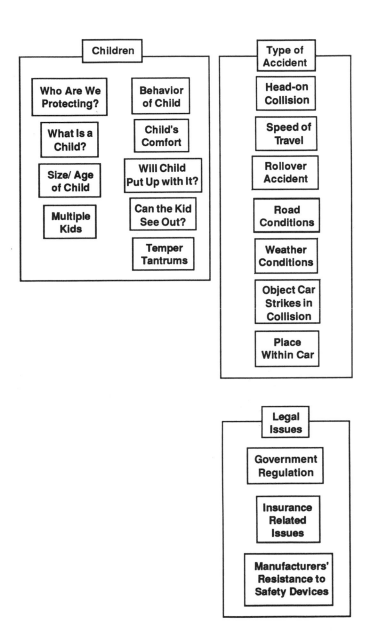

Children

- Who Are We Protecting?
- What Is a Child?
- Size/ Age of Child
- Multiple Kids
- Behavior of Child
- Child's Comfort
- Will Child Put Up with It?
- Can the Kid See Out?
- Temper Tantrums

Type of Accident

- Head-on Collision
- Speed of Travel
- Rollover Accident
- Road Conditions
- Weather Conditions
- Object Car Strikes in Collision
- Place Within Car

Legal Issues

- Government Regulation
- Insurance Related Issues
- Manufacturers' Resistance to Safety Devices

Figure 9
KJ METHOD/AFFINITY DIAGRAM
Expectations of GOAL/QPC

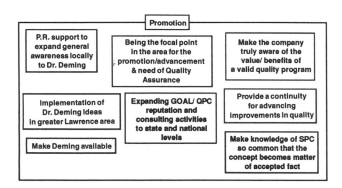

Promotion

P.R. support to expand general awareness locally to Dr. Deming

Being the focal point in the area for the promotion/advancement & need of Quality Assurance

Make the company truly aware of the value/ benefits of a valid quality program

Implementation of Dr. Deming ideas in greater Lawrence area

Expanding GOAL/ QPC reputation and consulting activities to state and national levels

Provide a continuity for advancing improvements in quality

Make Deming available

Make knowledge of SPC so common that the concept becomes matter of accepted fact

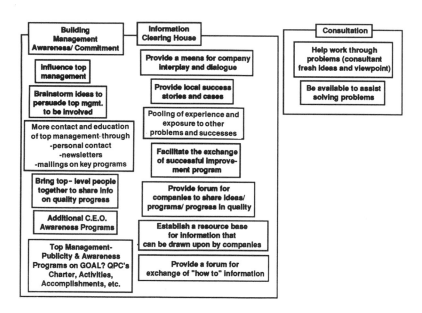

Building Management Awareness/ Commitment

Influence top management

Brainstorm ideas to persuade top mgmt. to be involved

More contact and education of top management through
-personal contact
-newsletters
-mailings on key programs

Bring top- level people together to share info on quality progress

Additional C.E.O. Awareness Programs

Top Management- Publicity & Awareness Programs on GOAL? QPC's Charter, Activities, Accomplishments, etc.

Information Clearing House

Provide a means for company interplay and dialogue

Provide local success stories and cases

Pooling of experience and exposure to other problems and successes

Facilitate the exchange of successful improve- ment program

Provide forum for companies to share ideas/ programs/ progress in quality

Establish a resource base for information that can be drawn upon by companies

Provide a forum for exchange of "how to" information

Consultation

Help work through problems (consultant fresh ideas and viewpoint)

Be available to assist solving problems

58

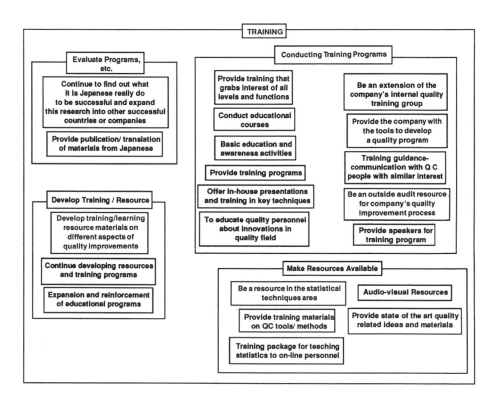

TRAINING

Evaluate Programs, etc.

Continue to find out what it is Japanese really do to be successful and expand this research into other successful countries or companies

Provide publication/ translation of materials from Japanese

Develop Training / Resource

Develop training/learning resource materials on different aspects of quality improvements

Continue developing resources and training programs

Expansion and reinforcement of educational programs

Conducting Training Programs

Provide training that grabs interest of all levels and functions

Conduct educational courses

Basic education and awareness activities

Provide training programs

Offer in-house presentations and training in key techniques

To educate quality personnel about innovations in quality field

Be an extension of the company's internal quality training group

Provide the company with the tools to develop a quality program

Training guidance- communication with Q C. people with similar interest

Be an outside audit resource for company's quality improvement process

Provide speakers for training program

Make Resources Available

Be a resource in the statistical techniques area

Provide training materials on QC tools/ methods

Training package for teaching statistics to on-line personnel

Audio-visual Resources

Provide state of the art quality related ideas and materials

Figure 10
KJ/METHOD/AFFINITY DIAGRAM
Issues in Implementing a Continuous
Improvement Process

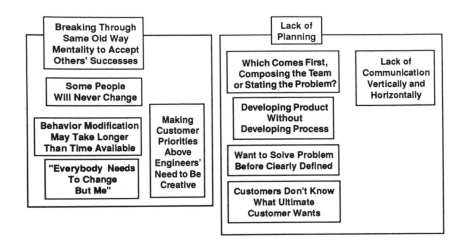

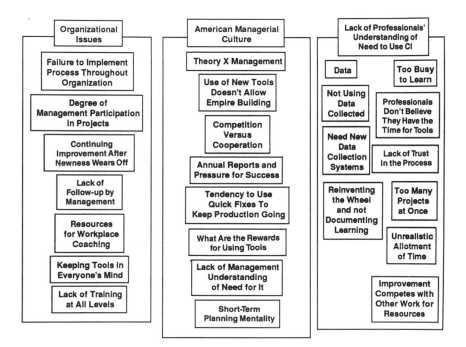

Organizational Issues	American Managerial Culture	Lack of Professionals' Understanding of Need to Use CI
Failure to Implement Process Throughout Organization	Theory X Management	Data
Degree of Management Participation in Projects	Use of New Tools Doesn't Allow Empire Building	Too Busy to Learn
Continuing Improvement After Newness Wears Off	Competition Versus Cooperation	Not Using Data Collected
Lack of Follow-up by Management	Annual Reports and Pressure for Success	Professionals Don't Believe They Have the Time for Tools
Resources for Workplace Coaching	Tendency to Use Quick Fixes To Keep Production Going	Need New Data Collection Systems
Keeping Tools in Everyone's Mind	What Are the Rewards for Using Tools	Lack of Trust in the Process
Lack of Training at All Levels	Lack of Management Understanding of Need for It	Reinventing the Wheel and not Documenting Learning
	Short-Term Planning Mentality	Too Many Projects at Once
		Unrealistic Allotment of Time
		Improvement Competes with Other Work for Resources

Figure 11
ISSUES INVOLVED IN CREATING THE IDEAL
CAR SAFETY SYSTEM FOR CHILDREN

Brainstorming of Group 1

Construction
Age of child
Upgradable?- Modularity
Freedom of movement
Crash resistance of car
Size of car
Ease of use
Cost of system
Play surface
Legal issues
Approval by standards body
Enforcement of code
Testing & validation requirements
Storage for small toys
Space taken up in vehicle
Multiple units
Comfort for occupant
Existing types-competition
Preciseness of specs
Product definition
Gain auto manufacturer support
Reliability
Durability
Ability to harass driver
Proximity to driver
Visibility to driver
Hardness
Brittleness
Ease of installation
Retrofit to older cars?
Require seat belt in car?
Air bags?
How to clean unit
Maintainability
Raw materials

Transportability
Emergency exit/ release
Fit 2-door or 4-door car
Research international market
Research international standards
Create market strategy
Customer training

Brainstorming of Group 2

Size/age of child
Size of car
Place within car
Speed of travel
Weather conditions
Head-on crash
What is a child?
Will child put up with it?
Temper tantrums
Road conditions
Rollovers
Type of accident
Design of safety devices
Design of vehicle
Ease of putting child in device
Ease of getting child out of device
Child's comfort
Mfgr.'s resistance to safety devices
Damage to car seat
Government regulations
Parental resistance
Color coordination of safety device
Consumer resistance
Insurance-related issues
Cost of safety device
Use of "Baby on Board" things
Can the kid see out?
Convertibles & head protection
Are doors locked?
Multiple safety devices
Childproof?
Portability
Easily cleaned
Emergency release
Multiple kids

Object car strikes in collision
Type of vehicle (car, truck, cycle)
What do consumers want?
Collapsible air bags
Safe loads for children
Waterbed car seats
Consumer attitudes
Load points on human body and
    safety padding locations
Deficiencies of current devices
Simplicity of instructions

3. The team should take the cards, mix them, and spread them out randomly on a large table.

4. The cards can be grouped by the team or assigned to the individual in the following way:

    a. Look for two cards that seem to be related in some way. Place those to one side. Look for other related cards to be placed together.

    b. Repeat this process until you have all possible cards placed in at most ten groupings. Do not force-fit single cards into groupings where they don't belong. These single cards ("loners") may form their own grouping or may never find a "home."

    Note 1: Do not refer to these as "categories." They are simply groupings of ideas that hang together. There is a difference between the two words.

    Note 2: It seems to be most effective to have everyone move the cards at will without talking. This forces team members not to get trapped in semantic battles.

    c. Look for a card in each grouping that captures the meaning of that group. This card is placed at the top of the grouping. If there is not such a card in the grouping, then one must be written. Remember that this card be simply and concisely written. Gather each grouping with the header card on the top.

5. Transfer the information on cards onto paper with lines around each grouping. Related clusters should be placed near each other with connecting lines. This is the first round of the KJ process and should be presented for additions, deletions, and modifications.

INTERRELATIONSHIP DIGRAPH

Definition.   *This tool takes a central idea, issue, or problem and maps out the logical or sequential links among related items. While still a very creative process, the Interrelationship Digraph begins to draw the logical connections that the KJ Method surfaces.*

In planning and problem solving, it is obviously not enough to just create an explosion of ideas. The KJ method allows some initial organized creative patterns to emerge but the Interrelationship Digraph (I.D.) lets **logical** patterns become apparent. This is based on the same principle that the Japanese frequently apply regarding the natural emergence of ideas. Therefore, an I.D. starts from a central concept and leads to the generation of large quantities of ideas, and finally the delineation of observed patterns. To some this may appear to be like reading tea leaves, but it works incredibly well. Like the KJ, the I.D. allows those unanticipated ideas and connections to rise to the surface.

When to Use the Interrelationship Digraph

We have found the I.D. to be exceptionally adaptable to specific operational issues as well as general organizational questions. For example, a classic use of the I.D. at Toyota focused on all of the factors involved in the establishment of a "billboard system" as part of their JIT program. On the other hand, it has also been used to deal with issues underlying the problem of getting top management support for TQC.

In summary, the I.D. should be used when:

a. An issue is sufficiently complex that the interrelationship between ideas is difficult to determine.

b. The correct sequencing of management actions is critical.

c. There is a feeling that the problem under discussion is only a symptom.

d. There is ample time to complete the required reiterative process.

Construction of an Interrelationship Digraph

As in the KJ diagram and the remainder of the tools, the aim is to have **the right people, with the right tools working on the right problems**. This means that the first step is to define the necessary blend of people among a group of six to eight individuals.

The construction steps are as follows:

1. **Clearly** define one statement that states the key issue under discussion.

Note: The source of this issue can vary. It may come from a problem that presents itself clearly. In this case, the I.D. would be the first step in the cycle rather than the KJ. The KJ is frequently used to generate the key issues to be explored in the I.D.

2. Record the issue/problem statement. It can be recorded by:

a. Placing it on the same type of card as is used in the KJ.

b. Writing it on a flip chart.

3. Place the statement in order to start the process in one of two patterns:

a. Centralized Pattern in which the statement is placed in the middle of the table or flip-chart paper with related ideas clustered around it.

b. Unidirectional Pattern in which the statement is placed to the extreme right or left of the table or flip-chart paper with related ideas posted on one side of it.

4. Generate the related issues/problems in the following ways:

a. Take the cards from a grouping under KJ and lay them out with the one that is most closely related to the problem statement placed next to it. Then lay out the rest of the cards in sequential or causal order.

b. Do wide open brainstorming, place the ideas on cards and cluster them around the Central Statement as in "a."

c. Do wide open brainstorming but directly onto the flip chart instead of cards. Proceed as in "a" or "b."

Note: Designate the central problem statement by placing it in a double lined egg-shaped enclosure. This double lining is referred to as "hatching."

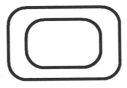

Note: The advantage of using cards is that they can be moved as the discussion progresses. The flip chart method is quicker but can become very messy if changes occur.

Note: When using the flip chart method designate all the related ideas by placing them in a single lined box.

5. Once all of the related idea statements are placed relative to the central problem statement fill in the causal arrows that indicate what leads to what. Look for possible relationships between each issue and every other issue.

Note: At this step you would look for patterns of arrows to determine what the key factors/causes are. For example, if one factor had seven arrows coming from it to other issues, while all others had three or fewer, then that would be a key factor. It would be designated by a double hatched box.

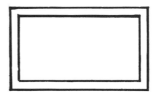

6. Copy the I.D. legibly and circulate to group members.

7. Meet for revisions and/or update and confirmation of the identified key factors.

8. As in the KJ, you may draw lines around groupings of related issues.

9. Prepare to use the identified key factors as the basis for the next tool, the **Tree Diagram**.

Figure 12
INTERRELATIONSHIP
DIGRAPH METHOD

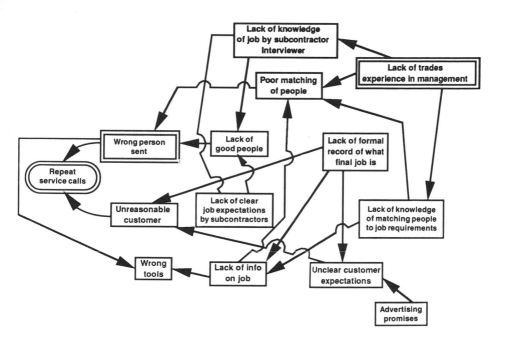

Figure 13

INTERRELATIONSHIP DIGRAPH INFLUENCING TOP MANAGEMENT

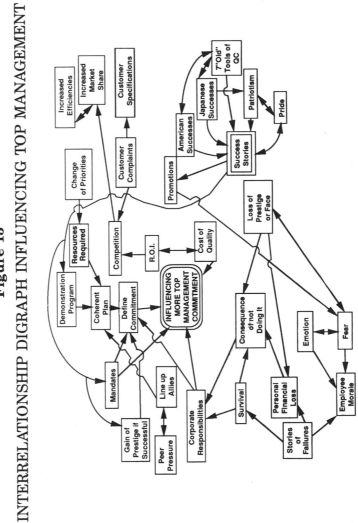

Figure 14
INTERRELATIONSHIP DIGRAPH

Lack of Professionals' Understanding of Need
Use Continuous Improvement Tools

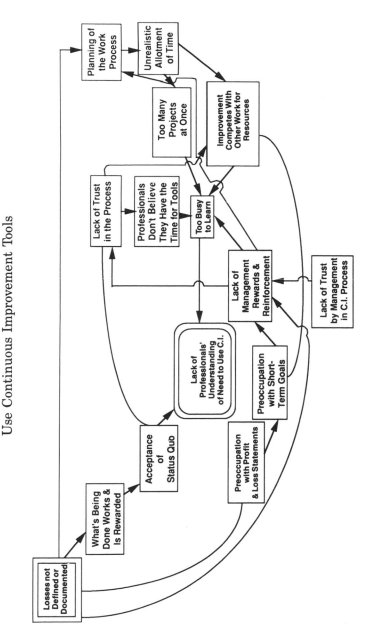

TREE DIAGRAM/SYSTEMS FLOW DIAGRAM

Definition. *This tool systematically maps out the full range of paths and tasks that need to be accomplished in order to achieve a primary goal and every related subgoal. In the original Japanese context, it describes the "methods" by which every "purpose" is to be achieved.*

In many ways, the KJ Method and Interrelationship Digraph force the key issues to the surface. The questions then become, "What is the sequence of tasks that need to be completed in order to best address that issue?" or "What are all of the factors which contribute to the existence of the key problem surfaced?" The Tree Diagram is appropriate for either question. Therefore, it can either be used as a cause-finding problem solver or a task-generating planning tool. In either use it brings the process from a broad level of concern to the lowest practical level of detail.

Another strong point is that it forces the user to examine the logical link between all of the interim tasks. This addresses the tendency of many managers to jump from the broad goal to details without examining what needs to happen in order for successful implementation to occur. It also rapidly uncovers gaps in logic or planning.

When to Use the Tree Diagram/Systems Flow Diagram

The Tree Diagram is indispensable when you require a thorough understanding of what needs to be accomplished, how it is to be achieved, and the relationships between these goals and methodologies.

It has been found to be most helpful in situations such as the following:

a. When you need to translate very ill-defined needs into operational characteristics. For example, a Tree Diagram would be helpful in converting a desire to have an "easy-to-use VCR" into every product characteristic that would contribute to this goal. It would also identify which characteristics can presently be controlled.

b. When you need to explore all the possible causes of a problem. This use is closest to the Cause & Effect Diagram (Fishbone Chart). For example, it could be used to uncover all of the reasons why top management may not support a continuous improvement effort.

c. When you need to identify the first task that must be accomplished when aiming for a broad organizational goal. For example, the Tree Diagram would be very helpful to the coordinator of Quality Improvement Programs who wants to know what is already being accomplished and where the key gaps exist.

d. When the issue under question has sufficient complexity and time available for solution. For example, a Tree Diagram would not be particularly helpful to decide how to deal with a product contamination problem that is shutting down your production line. It could be used to prevent it from re-occurring, but not in deciding on the stop-gap measures to be taken.

Note: In its most common usage the Tree Diagram conceptionally resembles the Cause & Effect Diagrams. We have found it to be easier to interpret because of its clear, linear layout. It also seems to create fewer "loose ends" than the C&E.

Construction of a Tree Diagram/Systems Flow Diagram

It has been shown that these tools are most powerful when used in combination, but they are also very effective when applied singly. With this in mind the following are the most widely used steps:

1. Agree upon one statement that clearly and simply states the core issue, problem, or goal. This statement may or may not come from a KJ Chart or Interrelationship Digraph.

Note: Unlike the KJ Method, the Tree Diagram becomes more effective as the issue is more clearly specified. This is important since the emphasis is on finding the logical and sequential links between ideas/tasks and not pure creativity.

2. Once the statement is agreed upon, the team must generate all of the possible tasks, methods, or causes related to that statement. These could follow three different formats:

a. Use the cards from the KJ Chart as a foundation. For example, you might take the ten to twenty cards that fall under one broad heading as a starting point.

b. Brainstorm all of the possible tasks/methods/causes and record them on a flip chart. These ideas could then be placed on individual cards or rearranged on the flip chart.

c. Brainstorm as in "b" but record directly onto cards for continued use.

Note: When brainstorming, continue to apply to each idea the question "In order to achieve _____ what must happen or exist?" Or "What has happened or what exists that causes _____?"

3. Evaluate and code all of the ideas with the following code:
  ○ Possible to carry out
  △ Need more information to see if possible
  × Impossible to carry out

Note: Code an idea to be impossible only after very careful consideration. "Impossible" must not be equated with "We've never done it before."

4. Construct the actual Tree Diagram by:

   a. Placing the central goal/issue card to the left of a flip chart or table. (In fact, the remainder of the instructions will assume that cards are being used, but the same steps would apply if the chart is drawn directly on the flip chart.)

   b. Ask the question, "What method or task do we need to complete in order to accomplish this goal or purpose?" Find the ideas on the cards or flip chart list that are the most closely related to that statement. These may also be viewed as those tasks that are the closest in terms of sequence or cause & effect.

   c. Place the ideas/tasks from "b" immediately to the right of the central issue card as you would in a family tree or organizational chart.

   d. The ideas/tasks from "c" now become the focal point. In other words, the question from "b" is repeated and the remaining cards are again sorted to be placed to the right as the next row in the Tree. This process is repeated until all of the cards or recorded ideas are exhausted.

   Note: If none of the cards answers the repeated question, create a new one and place it in the proper spot.

e. Review the entire Tree Diagram to ensure that there are no obvious gaps in sequence or logic. Check this by reviewing each path, starting at the most basic task to the extreme right. Ask of each idea/task, "If we do _____ will it lead to the accomplishment of this next idea/task?"

f. Review with other groups for relevant input and revise where needed.

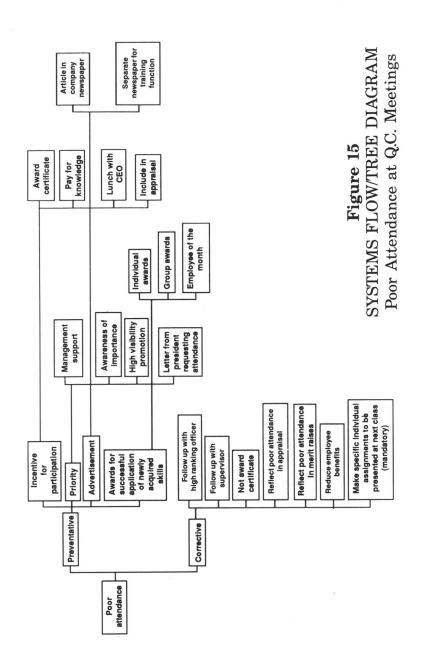

Figure 15
SYSTEMS FLOW/TREE DIAGRAM
Poor Attendance at Q.C. Meetings

77

Figure 16
TREE DIAGRAM
Daily Operational Changes Needed for a
Continuous Improvement Program

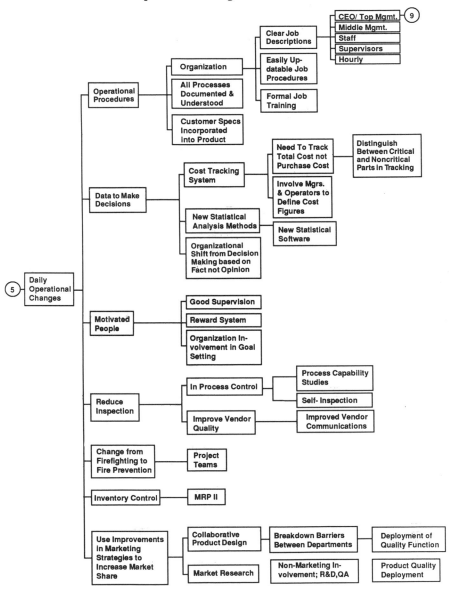

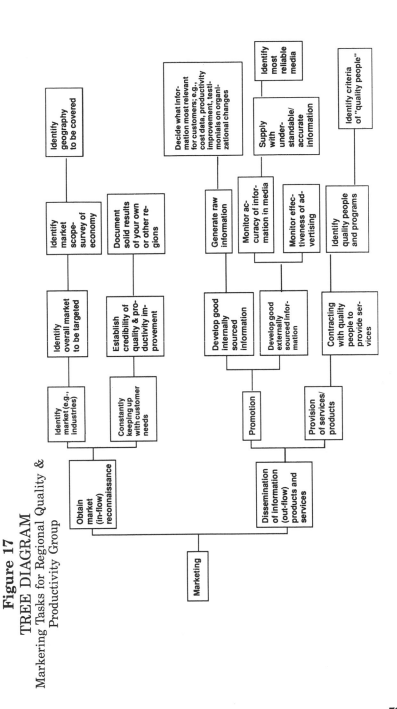

Figure 17
TREE DIAGRAM
Marketing Tasks for Regional Quality &
Productivity Group

79

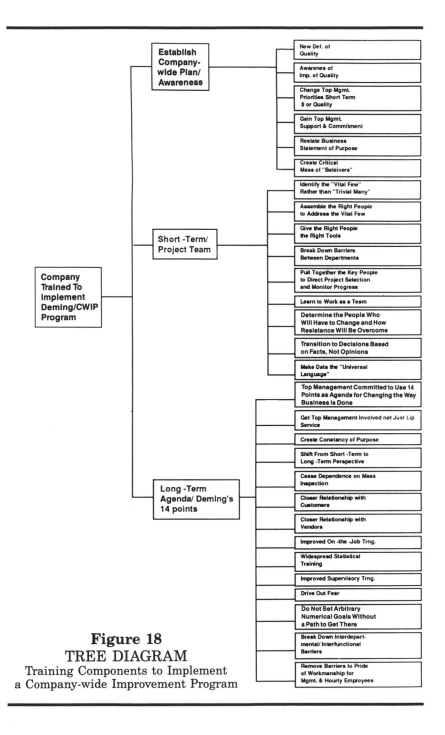

Figure 18
TREE DIAGRAM
Training Components to Implement
a Company-wide Improvement Program

Figure 19
TREE DIAGRAM
Ways to Deal with the Lack of Management Rewards &
Reinforcement for Continuous Improvement

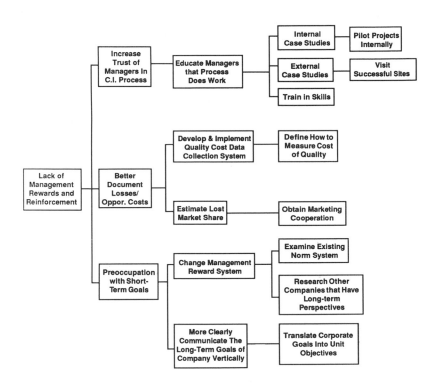

Figure 20
TREE DIAGRAM
Designing a Product to Meet Customers' Wants and Needs

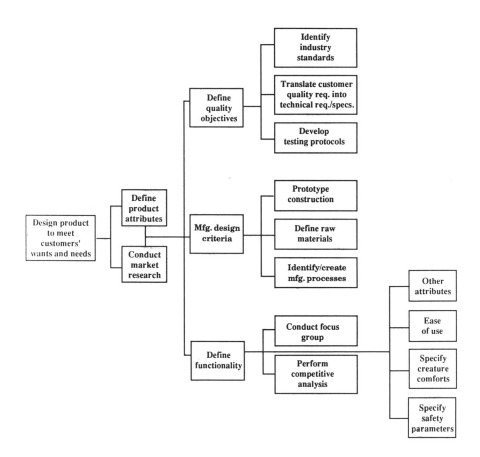

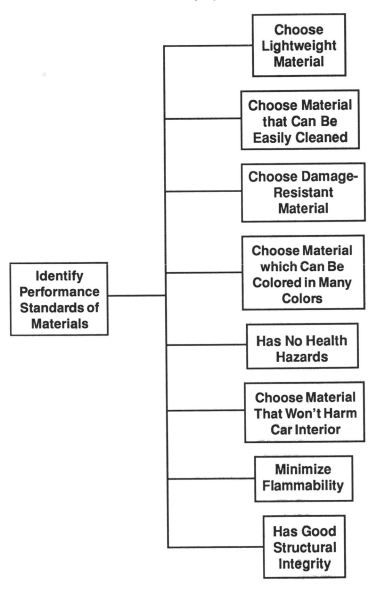

Figure 21
TREE DIAGRAM
Identifying Performance Standards of Materials
(Child Safety System)

MATRIX DIAGRAM

Definition.   *This tool organizes large groups of characteristics, functions, and tasks in such a way that logical connecting points between each one are graphically displayed. It also shows the importance of each connecting point relative to every other correlation.*

Of the tools discussed thus far (KJ Method, Interrelationship Digraph, Systems Flow/Tree Diagram), the Matrix Diagram has enjoyed the widest use. It is based on the principle that whenever a series of items are placed in a line (horizontal) and whenever a series of items are placed in a row (vertical), there will be intersecting points that indicate a relationship. Furthermore, the Matrix Diagram features highly visible symbols that indicate the strength of the relationship between the items that intersect at that point. Thus, the Matrix Diagram is very similar to the other tools in that new cumulative patterns of relationships emerge based on the interaction between individual items. Even in this most logical process, unforeseen patterns "just happen."

When to Use the Matrix Diagram

Because the Matrix Diagram has enjoyed the widest use of the New Tools, it has evolved into a number of forms. The key to successfully applying a Matrix Diagram is the choice of the right format matrix for the right situation. The following are the most commonly used matrix formats:

a. **L Shaped Matrix Diagram**

This is the most basic form of Matrix Diagram. In the L Shape, two interrelated groups of items are presented in line and row format. It is a simple two-dimensional representation that shows the intersection of related pairs of items as shown in Figure 22. It can be used to display relationships between items in countless operational areas such as administration, manufacturing, personnel, R&D, etc. In Figure 23 it is used to first identify all of the organizational tasks that need to be accomplished and how they should be allocated to individuals.

Note: It is doubly interesting if each person completes the matrix individually and then compares the coding with everyone in the work group.

Figure 22
L SHAPED MATRIX

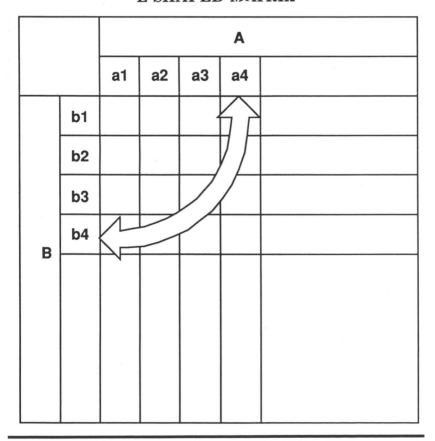

Figure 23
L SHAPED MATRIX
Allocation of Organizational Tasks

◎ Primary Responsibility
○ Secondary Responsibility
△ Communications/ Need to Know

+ Slightly More Emphasis

	Bob	Mike	Lee	Larry	Ann	Pat	Lynn	Board of Dir.	Other
Administration									
Payroll	◎								
Benefits	○	△	△	△	△	△	△	△	
Office Systems	○	○	◎			◎	△	○	
Computer Programs	○	△	◎			○	○		
Courses									
Update Mailing List	◎		○			◎	◎		
Select Courses to Be Offered	◎	◎	◎			△	△		△ Deming
Approve Course Content	◎	◎	○			△			△ Deming ○ Instructor
Prepare Brochures	○	○	○		◎	△			○ Instructor
Prepare Mailing	△		△			◎	○		
Hotel Arrangements	△	△	◎			△+	△		
Order Materials	△	△	◎			○	△		
Register People	△	△	△			◎	○		
Copy Materials	△	△	△			◎	○		
Prepare Packets	△	△	△			◎	○		
Room Set-Up	◎	◎	◎						
Post Receipts	△		◎				◎		
Prepare bills									
New Course Development									
Market Research	○	△	△			△			
Implementing Deming	◎	◎	△			△			
TQC	◎	○	△	○			○		

Figure 24

L SHAPED MATRIX
Shipping Problems

Department/ Individually/ Function / Problems	Customer Service	Quality	Production	Scheduling	Process Engineering	Shipping	Design Engineering
Missing Parts	△		○			◎	
Does Not Meet Specs	△	○	◎		○		△
Wrong Parts	△	○		○		◎	
Mis-labeled	△		◎			○	
Defective Parts	△	○	◎		○		△
Arrived Late	△		○	◎		○	
Too Many	△		○			◎	
Too Few	△		○			◎	
Shipping Damage	△					◎	○
Wrong Part Ordered	◎						
Customer No Longer Needs Part	◎						
Cannot Process Part	△	○	○		◎		◎

◎ Primary Responsibility
○ Secondary Responsibility
△ Communications/Receive Reports

Figure 24 shows yet another application to an all-too-common problem: Shipping Problems. By brainstorming all of the possible reasons for shipping problems it is very clear that the "shipping" problem does not only rest with the shipping department. The matrix forces the participants to also develop the list of all related departments. The interrelationships between these two sets of items point to the pattern of responsibility for solution to the problems.

Figure 25
L SHAPED MATRIX
Designing a Product to Meet Customer Needs (Child Safety System)

	Marketing	Development R&D	Quality	Manufacturing	Process Engineering	Exec.	Purchasing
Identify Industry Standards	○	◎	◎	△	△		△
Translate Customer Req. Into Specs	△	◎	◎	○	○		
Develop Testing Protocols		◎	◎	△			
Prototype Construction	△	◎	◎	◎	○+	△	△
Define Raw Materials	△	◎	○	△	○		◎
Create Manufacturing Process	△	○	△+	◎	◎	△	○
Conduct Focus Groups	◎	○	○	○			
Perform Competitive Analysis	◎	△	○	△	△	△	△

△ Keep Informed   ○ Secondary Responsibility   ◎ Primary Responsibility

Figure 26
L SHAPED MATRIX

Materials

	Metals					Plastics				Fabrics					Foam		Natural	
Identify Performance Standards Of Materials	S.STEEL	C.ALUM	M.STEEL	C.ALLOY		RUBBER	URETHANE	RUBBER	PLASTS	CWOOD	WOOL	NYLON	POLYEST	ACETATE	CBFOAM	HBFOAM	WOOD	LEATHER
Choose Lightweight Material	X	O	X	O	△	O	O	O	O	O	O	O	O	O	O	O	△	O
Choose Material That Can Be Easily Cleaned	O	X	X	O	O	O	O	O	O	X	X	O	O	O	X	X	X	X
Choose Damage Resistant Material	O	X	O	O	△	O	X	O	X	X	X	O	X	X	X	X	X	X
Has No Health Hazards	O	O	O	O	X	△	△	△	△	O	O	△	△	△ (Toxic Fumes?)	△	△	△	O
Choose Material That Won't Harm Interior of Car	X	O	X	O	O	O	O	O	O	O	O	O	O	O	O	O	O	O
Minimize Flammability	O	O	O	O	O	△	△	△	△	△	O	△	△	△	△	△	△	O
Has Good Structural Integrity	O	X	X	O	△	O	X	O	X	X	X	O	△	△	△	△	O	O
	+2+2-0-8+1					+6+2-6-2				+1+2-6-3-3					+1+1		-3+4	

(+1) - has
(-1) x - does not have
(0) - ?

Figure 27

BRAINSTORMING FOR
TYPE OF SAFETY DEVICE

Water- supported system
Adjustable inflation system
Air bags
Padded area for older children
Modified leg support system
Reclining system
Car seat- inflatable padding
Car seat- with sound system
Car seat- child accessible storage system
Modular car seat/ frame/ padding/ Acc.
Adjustable harness- impact activated (restraint)   Group 1
                                                                        _____

                                                                        Group 2
Cage   added
Padded cage
Suit
Seat belt
Shoulder belt
Foam interior of car

Figure 28
L SHAPED MATRIX
Comparison of Customer Requirements and Child Safety
System Design Options

		Heated Waterbed System	Adjustable Inflatable System	(Assume seat) Air Bags	Playroom w/ Air Bags	Playroom w/ Padding	Car Seat w/ Modif. Leg Support	Car Seat w/ Reclining	Car Seat w/ Inflatable Padding	Car Seat w/ Sound System	Car Seat w/ Storage (Child Access)	Modular Car Seat by Component	Impact Activated Adj. Harness
Age of Child 0-1	Needs Head Support	◎	◎	X	X	X	○	◎	◎	○	○	◎	X
	Needs Soft Padding	◎	◎	X	○	○	○	○	◎	○	○	◎	X
	Back Should Absorb Shock	◎	○	X	X	X	◎	◎	◎	◎	◎	◎	X
Age 3	Needs Attention Occupiers	X	X	X	◎	◎	X	○	X	◎	○	◎	X
	Child Proof Restraint	X	X	◎	◎	◎	○	○	○	○	○	○	◎
Stds. " " " "	Emergency Release	X	○	◎	◎	◎	◎	◎	◎	◎	◎	◎	◎
	Flame Retardant	◎	○	◎	◎	◎	◎	◎	◎	◎	◎	◎	◎
	Withstand Klb/In²	◎	○	◎	◎	◎	◎	◎	◎	◎	◎	◎	◎
	Immobile During Use	◎	◎	◎	◎	◎	○	○	○	○	○	X	◎
	Not Toxic	◎	◎	◎	◎	◎	◎	◎	◎	◎	◎	◎	◎
Comfort " " "	Mobility of Head	◎	○	◎	◎	◎	○	○	○	○	○	○	◎
	Mobility of Arms	◎	◎	◎	◎	◎	◎	◎	◎	◎	◎	◎	◎
	Legs Supported	◎	◎	X	◎	◎	◎	◎	○	○	○	◎	X
Adjustable "	Surface Cool in Summer & warm in Winter	○	○	X	○	◎	○	○	◎	○	○	○	X
	Play Surface	X	X	X	◎	◎	X	X	X	X	X	○	X
	Grow with Child	X	○	◎	◎	◎	X	X	◎	◎	X	◎	◎
	Fits All Cars	○	○	◎	X	X	◎	◎	◎	◎	◎	◎	X

◎ Meets completely
○ Meets somewhat
X Does not meet

Figure 29
L SHAPED MATRIX
Comparison of Potential Construction Materials and Child Safety System Design Option

	Mg Al	Rt Plas	Nylon	Foam	Leather
Recycling System	◉	○	○	○	X
Cage	◉	X	◉	◉	X
Car Seat w/ storage sound, padding	◉	○	○	◉	X
Adjustable Restrained Impact Activated (Harness)	○	○	◉	X	○

◉ Ideal
○ Adequate
X Not Appropriate

b. T Shaped Matrix

The T Shaped Matrix is nothing more than the combination of two L Shaped Matrix Diagrams. As can be seen in Figure 30, it is based on the premise that two separate sets of items both are related to a third set. Therefore, A items are somehow related to both B and C items.

Figure 31 shows one application. In this case, it shows the relationship between a set of courses in a curriculum and two important sets of considerations: Who should do the training for each course? And, which would be the most appropriate functions to attend each of the courses?

It has also been widely used to develop new materials by simultaneously relating different alternative materials to two sets of desirable properties.

c. Y Shaped Matrix

The Y Shaped Matrix simply allows the user to combine and compare three sets of items to each other. As in Figure 32, it is clear that you can now determine the interaction between items in Group A with those in Group B, as well as Group B with Group C and Group C with Group A.

This is invaluable when comparing product characteristics, etc.

d. X Shaped Matrix

The X Shaped Matrix is a format that is rarely used. It shows the interaction between four sets of items. As in Figure 33, it relates graphically A&B, B&C, C&D, and D&A. It is available, but its use is not well documented.

Figure 30
T SHAPED MATRIX

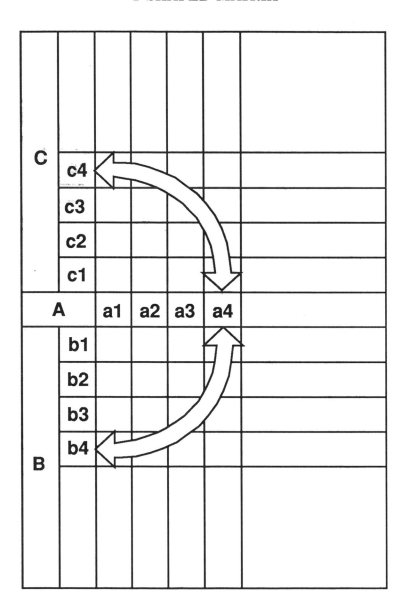

Figure 31

T-MATRIX DIAGRAM ON COMPANY-WIDE TRAINING

Who Trains?

- Human Resource Dept.
- Managers
- Operators
- Consultants
- Production Operator
- Craft Foremen
- GLSPC Coordinator
- Plant SPC Coordinator
- University
- Technology Specialists
- Engineers

*Need to Tailor to Groups

Courses

X = Full
O = Overview

Column headers (Courses): SQC | 7 Old Tools | 7 New Tools | Reliability | Design Review | QC Basics | QCC Facilitator | Diagnostic Tools | Problem Solving | Communication Skills | Organize for Quality | Design of Experiment | Company Mission | Quality Planning | Just In Time | New Superv. Training | Comp. Tot. Q. Mgt. Syst. | Group Dynamics Skills | SQC Course/ Execs.

Who Attends?

- Executives
- Top Mgmt.
- Middle Mgmt.
- Prod. Supervisors
- Supp. Func. Mgrs.
- Staff
- Marketing
- Sales
- Engineers
- Clerical
- Prod. Worker
- Qual. Professional
- Project Team
- Emp. Involv. Teams
- Suppliers
- Maintenance

Figure 32
Y SHAPED MATRIX

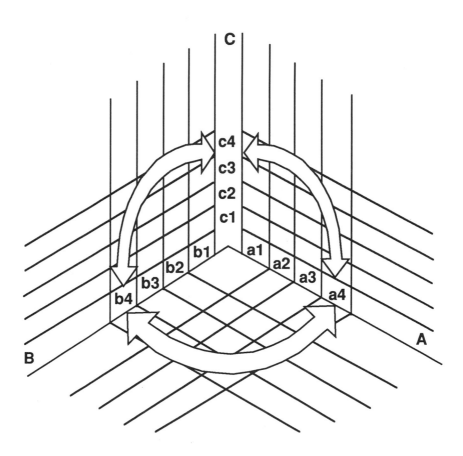

Figure 33
X SHAPED MATRIX

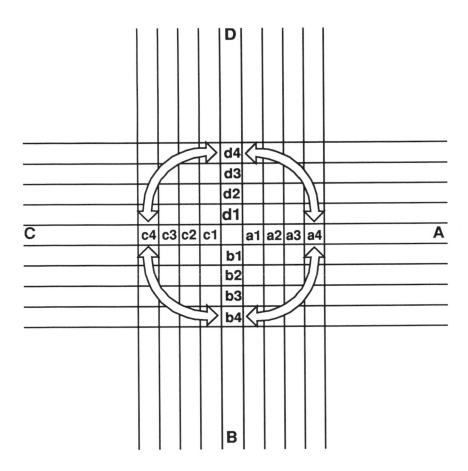

97

e. C Shaped Matrix

The C Shaped Matrix (or Cubic type Matrix) makes it possible to visually represent the intersection of three inter-related sets of items.

Other Matrix Diagram formats allow you to show the relationship between three or even four sets of items. In effect, however, they only compare two sets of items at a time with any connections with a third set only by inference. In other words, A is connected with B, B is connected with C, so it can be inferred that A is related in some way to C.

The advantage of the C Shaped Matrix is that it can graphically display the connection between A, B, and C directly as one converging point.

Figure 34 shows a C Shaped Matrix displaying the interaction between Layout, Software, and Hardware items. In this case, there is a strong connection between 4 under Layout (Select Software), 13 under Hardware (Measure Raw Materials), and 3 under Software (Be Flexible Against Changes).

f. Combination Matrix/Tree Diagram

As important as selecting the right format matrix is generating the most complete set of items practically possible. The Tree Diagram is widely used to generate the tasks, ideas, and/or characteristics that form one or more sides of the matrix.

Figure 35 shows two Tree Diagrams that have been merged into a simple L Shaped Matrix. Even more common than this is using a Tree Diagram to create a set of tasks to be accomplished (vertical axis of matrix) and merge them in an L Shaped Matrix with all of the departments/functions (horizontal axis). The degrees of responsibility for each task can then be clearly allocated and indicated.

Figure 34
C SHAPED MATRIX
Characteristics of Measuring System for Granules

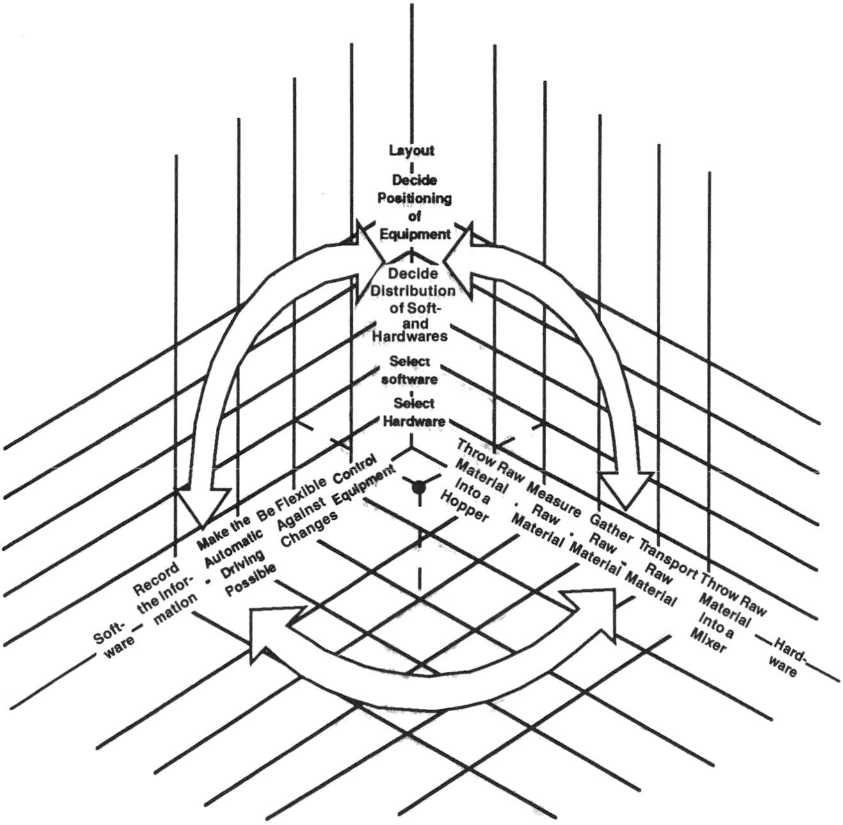

Figure 35
COMBINATION OF MATRIX DIAGRAM & TREE DIAGRAM

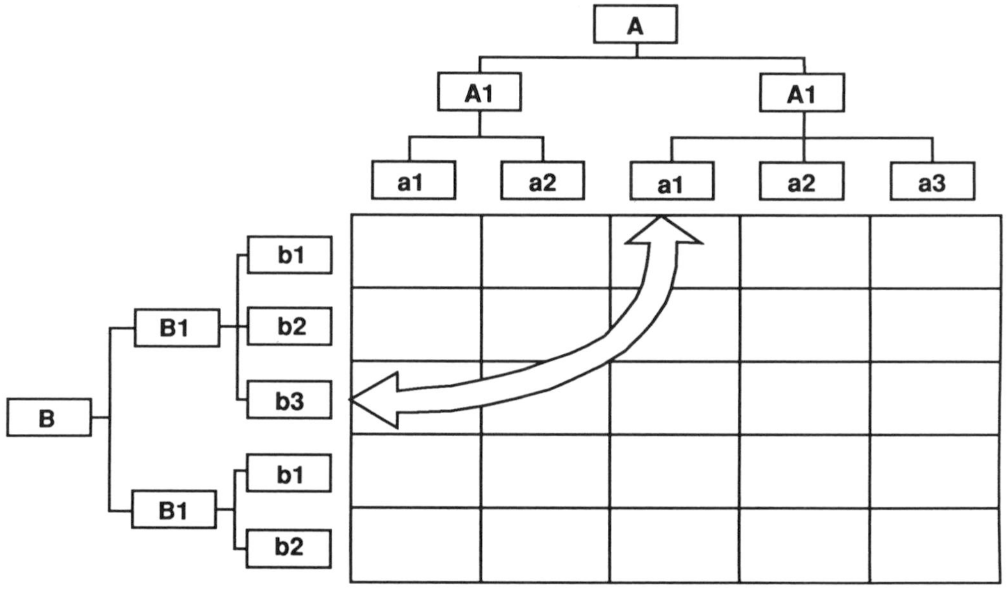

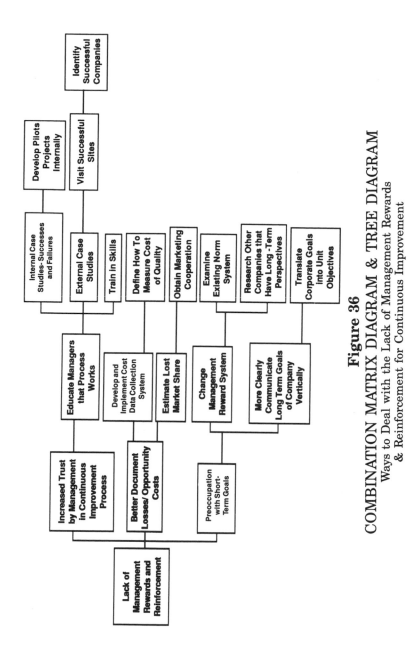

Figure 36

COMBINATION MATRIX DIAGRAM & TREE DIAGRAM

Ways to Deal with the Lack of Management Rewards
& Reinforcement for Continuous Improvement

Function	1	2	3	4	5	6	7	8
Purchasing				◉		△		◉
Personnel	△			○		◉	◉	◉
Marketing		◉		○△	◉	△		◉
Quality	◉	○		◉		△		◉
Training	△			△	◉	◉	◉	◉
Executive Committee	△	○		△		◉		◉
Accounting	○			◉		△		◉
Line Supervision	◉			○		△		◉
Sales		◉		○	◉	△		◉
(Q) MIS	○			◉		△	◉	◉
Statisticians	○			△○		○		◉
Consultants		○					◉	
Customers		○						
Engineering	◉			◉	◉	△		◉

Continuation of Figure 36

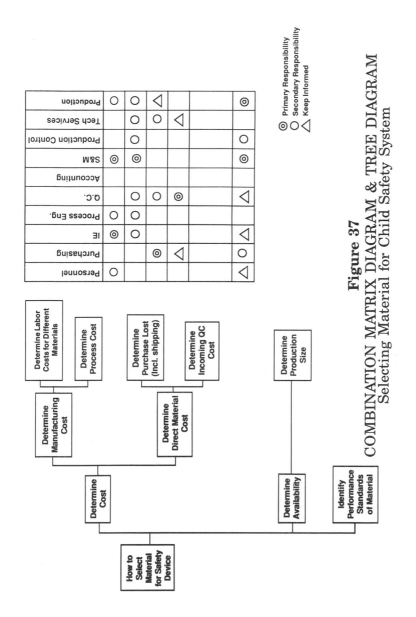

Figure 37

COMBINATION MATRIX DIAGRAM & TREE DIAGRAM
Selecting Material for Child Safety System

◎ Primary Responsibility
○ Secondary Responsibility
△ Keep Informed

Construction of a Matrix Diagram

The process of constructing any of the various format Matrix Diagrams is very straightforward. It is as follows:

1. Generate the two, three, or four sets of items that will be compared in the appropriate matrix.

   Note: These often emerge from the last row of detail in a Tree Diagram. This is the most effective method, but the matrix has proved helpful when based upon brainstormed items from a knowledgeable team.

2. Determine the proper matrix format.

   Note: The choice of sets of items to compare is based on an educated guess and experience. It is trial and error. Don't be afraid to abandon or modify a line of reasoning.

3. Place the sets of items so as to form the axes of the matrix. If these items come from one or more Tree Diagrams, you can simply tape the cards (if used) on a flip chart pad. Otherwise you can simply record them directly on the pad. Finally, draw the lines which will form the boxes in which the appropriate relationship symbols will be placed (see step 4).

4. Decide on the relationship symbols to be used. The following are the most common, but use your imagination.

   • Function Responsibility Chart
      ◉ Primary Responsibility
      ○ Secondary Responsibility
      △ Should Receive Information

   • Quality Characteristics Chart
      A Most Critical
      B More Critical
      C Critical

- Product Testing Chart
  - Test In Process
  - ○ Test Scheduled
  - × Test & Evaluation Possible

Note: Regardless of which symbols you choose to use, be sure to include a legend that prominently displays the relationship symbols used and their meaning.

MATRIX DATA ANALYSIS

Definition.   *To take data displayed in a Matrix Diagram and arrange it so that it can be more easily viewed and reveal the true strength of the relationship between variables.*

When to Use Matrix Data Analysis

Matrix Data Analysis is used primarily for market research, planning, and development of new products and process analysis and is used to determine the representative characteristics of each variable being examined. For example, what are the demographic characteristics of groups of people who like or dislike certain foods? What are the representative characteristics of a new cloth given an array of possible end uses?

Construction of a Matrix Data Analysis Chart

1. In order to find the "representative characteristics" of a product or consumer, use the "Principal Component Analysis Method." It is a formula to calculate mathematically the impact that a factor has on the process.

2. Compare data among evaluation groups showing how much of the intergroup variation is due to a particular characteristic of that group (or combination thereof).

3. Calculate the cumulative contribution rates of the principle components to the overall ratings, e.g., sex, age, and occupation accounted for 75% of the variability in the rating.

4. Display the distribution of results graphically in a four quadrant graph.

Figure 38
MATRIX DATA ANALYSIS EXAMPLE

Distribution of Different Food Tastes

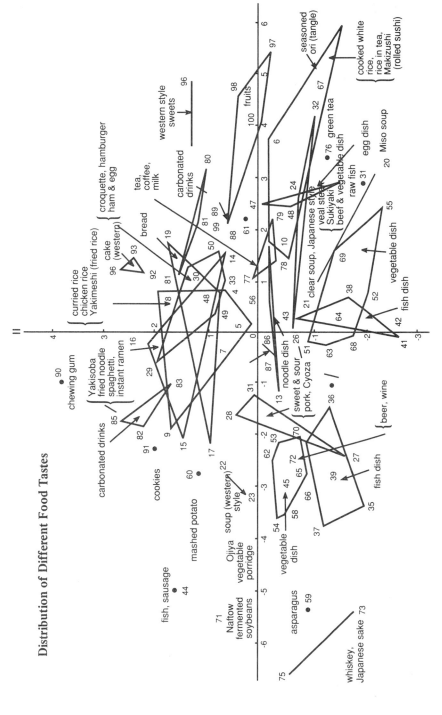

PROCESS DECISION PROGRAM CHART (PDPC)

Definition.   *Process Decision Program Chart (PDPC) is a method which maps out every conceivable event and contingency that can occur when moving from a problem statement to possible solutions. This tool is used to plan each possible chain of events that needs to occur when the problem or goal is an unfamiliar one.*

The underlying principle behind the PDPC is that the path toward virtually any goal is filled with uncertainty and an imperfect environment. If this weren't true, we would have a Deming "sequence" like the following:

PLAN ——→ DO

Reality makes the Deming **Cycle** a necessity.

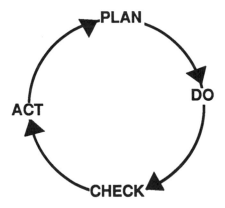

PDPC anticipates the unexpected and, in a sense, attempts to "short circuit" the cycle so that the "check" takes place during a dry-run of the process. The beauty of PDPC is that it not only tries to anticipate deviations, but also to develop countermeasures that will either:

a. prevent the deviation from occurring or

b. be in place in case the deviation occurs.

The first option is ideal in that it is truly preventive. However, we live in a world of limited resources. In allocating these resources we have to often "play the odds" as to the chance of X, Y, or Z happening. Given that fact, the next best thing is to have a contingency plan in place when a case occurs that we were "betting against." PDPC provides a structure to go in either direction.

When to Use a PDPC

An ideal use of a PDPC would be as follows:

A scientist's goal is to explore the core of the earth to determine its composition. Her plan is to dig a tunnel four miles deep. It has never been done, she doesn't know how long it will take, but she still has to make a funding proposal. The questions are then: How do you describe all of the possible paths to achieve this goal? How do you know what some of the obstacles will be? How can you prevent these possibilities from becoming realities? If obstacles do occur, how do you react in a timely way so as to avoid going back to "square one"? A cost estimate will be possible only if these questions have been answered. How can this be done systematically? Simple. . . PDPC!

It is obviously not so simple, but it certainly provides a methodical structure that can prevent details from slipping between the cracks.

Note: PDPC is most like the Tree Diagram in structure and aim since both deal with possible patterns of methods and plans. In the same vein it is closely tied with methods in reliability engineering such as Failure Mode & Effect Analysis (FMEA) and Fault Tree Analysis (FTA). The prime difference between these two formats is that FMEA starts from the smallest detail (sub-system) and assesses the probability of failure at any step. Also, it determines the cumulative impact on the end goal. FTA, on the other hand, starts with an undesirable result and then traces it back sequentially looking for the cause. PDPC is enjoying widespread use in particular because of the stress on product liability.

Construction of a Process Decision Program Chart (PDPC)

Even though PDPC is a methodical process, it also has few guidelines in terms of the process and finished product. The most important thing to keep in mind is that you must get to the point where deviations and contingencies are **clearly** indicated. This must be true at every level of detail in the chart.

Note 1: The source of the goal statement that starts the PDPC process often emerges from tools such as the KJ Interrelationship Digraph or even the Tree Diagram. As applies to all the other tools, PDPC can also be used effectively on its own.

Note 2: One word of caution. EXPLOSION! This is how some users have described PDPC. The creation of possible paths and countermeasures can multiply the complexity of the chart tremendously. Don't let it overwhelm you. Break the material into bite-size pieces, develop each piece, and then reassemble the final product.

The following seems to be the most workable approach:

1. Follow the instructions for the Tree Diagram through the end.

2. Take one branch of the Tree Diagram (starting from the "purpose" in the row to the immediate right of the "ultimate goal/purpose") and ask the questions: What could go wrong at this step? or, What other path could this step take?

Note: It seems to be easier if the items in that original branch are on cards since they can be moved easily. This is important since you are inserting problems and countermeasures into an existing sequence.

3. Answer the questions in "b" by branching off the original path.

4. Off to the side of that step, list actions/countermeasures that could be done. These are normally enclosed in "clouds" similar to cartoon captions and attached to that problem statement.

5. Continue the process until that original branch is exhausted.

6. Repeat "b" through "e" on the next most important tree branch, etc.

7. Assemble the individual branches into a final PDPC and review with the proper team of people and adjust as needed.

Figure 39
PROCESS DECISION PROGRAM CHART (PDPC)
Possible Product Defects That Could Occur in a
25 MPH Rear End Collision (Modular Car Seat Option)
Group 1

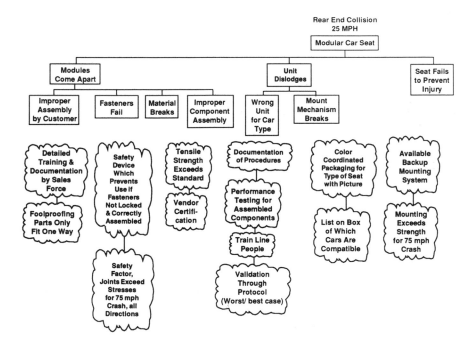

Figure 40
PROCESS DECISION PROGRAM CHART (PDPC)
Possible Product Defects That Could Occur in a
25 MPH Rear End Collision (Air Bags Option)
Group 2

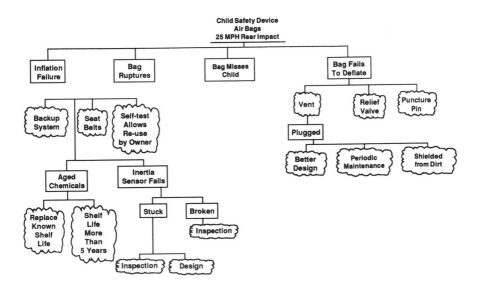

Figure 41
PROCESS DECISION PROGRAM CHART (PDPC)

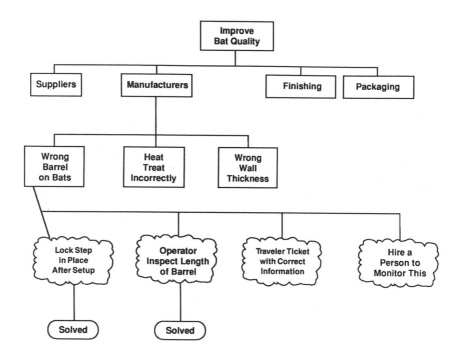

ARROW DIAGRAM

Definition. *This tool is used to plan the most appropriate schedule for any task and to control it effectively during its progress. This is closely related to the CPM and PERT Diagram methods. This is used when the task at hand is a familiar one with subtasks that are of a known duration.*

The Arrow Diagram is one tool that is certainly not Japanese. It is based on the Program Evaluation and Review Technique (PERT), which was developed in the U.S. after WWII to speed the development of the Polaris program. What the Arrow Diagram does is to remove some of the "black box magic" from the traditional PERT process. This is consistent with the general approach that the key to Japanese success is their ability to take previously available tools and make them accessible to the larger population. So, instead of industrial, manufacturing, and design engineers papering their walls with PERT charts (which they have), they can be used as a daily tool throughout the organization.

When to Use the Arrow Diagram

The most important criterion (and perhaps the only meaningful one) is that the subtasks, their sequencing, and duration must be well known. If this is not the case, then the construction of the Arrow Diagram can become a very frustrating experience. When the timing of the actual events is so different from the Arrow Diagram, people dismiss it as a nuisance, never to be used again (voluntarily).

When there is a lack of history about a process, the PDPC is likely to be a much more helpful tool.

Note: Don't be afraid to admit that you don't know everything there is to know about a process. It is better to decide on the proper tasks and sequencing than to pretend that you have a handle on the scheduling dimension.

Obviously, there are many processes that do have a well documented history. Therefore, the Arrow Diagram has enjoyed widespread use in such areas as:

- New Product Development
- Construction Projects
- Marketing Plans
- Complex Negotiations

Construction of an Arrow Diagram

As usual, a successful process is based on having complete input from the right sources. It's possible that one person could have all of the needed information for structuring an Arrow Diagram, but it would be highly unlikely. Therefore, assembling a team of the right people would be the usual first step. This team would follow the steps listed below:

1. The team would generate and record **all** of the necessary tasks to complete the project.

 Note 1: It is strongly recommended that these tasks be written simply and clearly on cards (about business card size or slightly narrower). This is essential for moving the cards before the final lines and arrows are drawn. Expect to generate 50-100 such cards.

 Note 2: For on-the-job cards, be sure to write the task to be completed only in the top half of the card. Draw a line under the task, thereby dividing the card in half. The length of time to complete that task will be filled in this space later.

2. Scatter the completed job cards and judge the interrelationship between jobs. Determine the relationship among the cards, e.g., what precedes, follows, or is simultaneous to each job and place them in the proper flow. Delete duplication and add new cards if jobs are overlooked.

3. Decide on the positions of the cards by finding the path with the most job cards in a series. Leave space between the cards so that "nodes" can be added later. These are the symbols that show the beginning and end of a task or event. Draw these in only when the various paths have been determined.

4. Find the cards whose path parallels the first path, then the path that parallels that one, etc.

5. Once these paths are finalized, write in the nodes, number them, and add arrows between tasks in each path as well as those linking each path to the other.

   Note: Each task/job is made up of two nodes. The task that begins with node #1 and ends with node #2 is task 1,2.

6. Study **carefully** the number of days, hours, weeks, etc. for each task and complete each job card.

7. Based on #5, calculate the earliest and latest start time for each node.

   Note 1: This is critical if you are to calculate the Critical Path (as in CPM), which is the longest cumulative time that the tasks require. This is therefore the shortest time that one could expect the final tasks to be completed.

   Note 2: The earliest and latest start times should be calculated using the following formulas:

a. 1) **Earliest Node Time**

   Suppose there is a job that starts from the node i. "Earliest node time" is the day when the job can be started. No sooner than that date. And it is expressed as t_i^E.

2) Latest Node Time

Suppose there is a job that ends at the node i. "Latest node time" is the day when the job must be finished. It is expressed as t_i^L. t_i^E and t_i^L will be written near the node.

b. How to Calculate Earliest Mode Time

Here is how to calculate earliest node time:

1) Earliest node time of the starting point (node 1) in the Arrow Diagram is O, i.e., $t_j^E = 0$.

2) When one job has a latter node j, its earliest node date t_i^E can be obtained using the following equation:

$$t_j^E = t_i^E + D_{ij}$$

Where t_i^E is earliest node time of starting node i of node j D_{ij} is the necessary days of the job (i,j).

3) When there are two jobs (or more) using the node j as their latter node, its earliest node time t_j^E can be obtained using the following equation:

$$t_j^E = \max (t_i^E + D_{ij})$$

c. How to Calculate Latest Node Time

1) Latest node time of the very last point (node n) in the Arrow Diagram has the same value as earliest node time of that node, i.e.,

$$t_n^L = t_n^E$$

2) When there is one job using node i as the starting node, its latest node time t_i^L can be obtained using the follow equation:

$$t_i^L = t_j^T - D_{ij}$$

Where t_j^L is the latest node time of following node j to the node i, D_{ij} is the necessary days for the job (i,j).

3) When there are more than two jobs which use node i as a preceding node, its latest node time t_i^L can be obtained using the following equation:

$$t_i^L = \min (t_j^L - D_{ij})$$

d. **Relationship between t_i^E and latest node time t_i^L at the same node.**

$$t_i^E = t_i^L$$

There is the following relationship between earliest node time t_i^E and latest node time t_i^L at the same node.

$$t_i^E = t_i^L$$

e. **Critical Path**
Critical path is the longest path from the starting point to the finishing point on the Arrow Diagram. It is the series of jobs important to the schedule control. Critical path should be as shown in the following equation:

$$t_i^E = t_i^L$$

Along the path, t_i^E and t_i^L should be posted, and the path should be shown in heavy directional line in the Arrow Diagram.

f. Slack

Slack means the margin time at the node i. The difference between latest node time and earliest node time will be expressed as SL_i.

$$SL_i = t_i^L - t_i^E$$

This slack is a general idea of a margin in schedule to be used when earliest node time and latest node time are calculated. In the case of a strict schedule control, you must calculate the total margin or free margin which will be mentioned later.

g. Example of Calculation

This is the example of the calculation of earliest and latest node times. The heavy line is the critical path.

Figure 42
CALCULATION OF NODE TIME

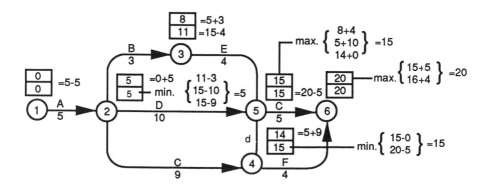

Symbols

1. **Event, node:** These are the beginning and the finish of a job, and they are the connecting points to other jobs.

2. **Job, activity:** This is the element that needs a length of time.

3. **Dummy:** This is the element that shows the interrelationship between jobs, but needs no time.

4. **Numbers for nodes:** These are the numbers at the nodes, events and they show which job it means, or in what order it places.

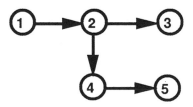

Bibliography

ARTICLES ON QFD

Cohen, Louis. "Quality Function Deployment: An Application Perspective from Digital Equipment Corporation." *National Productivity Review,* Summer 1988.

De Vera, Dennis, et al. "An Automotive Case Study." *Quality Progress,* June 1988.

Feigenbaum, Armand V. "Total Quality Leadership." *Quality,* April 1986.

Fortuna, Ronald M. "Beyond Quality: Taking SPC Upstream." *Quality Progress,* June 1988.

Hauser, John R., and Clausing, Don. "The House of Quality." *Harvard Business Review,* May-June 1988.

Hauser, John R., and Klein, Robert L. "Without Good Research Quality Is Shot in Dark." *Marketing News,* January 4, 1988.

International TechneGroup, "QFD Enhanced by ITI Software." *Metalworking News,* February 6, 1989.

Kenny, Andrew A. "A New Paradigm for Quality Assurance." *Quality Progress,* June 1988.

King, Robert. "Listening to the Voice of the Customer: Using the Quality Function Deployment System." *National Productivity Review,* Summer 1987.

Kogure, Masao, and Akao, Yoji. "Quality Function Deployment and CWQC in Japan." *Quality Progress,* October 1983.

McElroy, John. "For Whom Are We Building Cars?" *Automotive Industries,* June 1987.

McElroy, John. "QFD: Building the House of Quality." *Automotive Industries,* January 1989.

Makabe, Hajime, and Miyakawa, Masami. "A Positive Study on Reliability Management and Its Characteristic Aspect in Quality Control Activities in Japan." *International Conference on Quality in Tokyo,* 1987.

Newman, Richard G. "QFD Involves Buyers/Suppliers." *Purchasing World,* October 1988.

Pugh, S. "Concept Selection–The Method That Works." Proceedings of ICED, Rome, March 1981, ADK 5, paper M3/16, pp. 497-506.

Ross, Phillip J. "The Role of Taguchi Methods and Design of Experiments in QFD." *Quality Progress,* June 1988.

Sullivan, Lawrence P. "Quality Function Deployment." *Quality Progress,* June 1986.

Sullivan, Lawrence P. "Policy Management Through Quality Function Deployment." *Quality Progress,* June 1988.

BOOKS ON QFD

Eureka, William, and Ryan, Nancy. *The Customer-Driven Company.* Dearborn, Michigan: ASI Press, 1988.

King, Robert. *Better Designs in Half the Time–Implementing QFD in America.* Methuen, Massachusetts: GOAL/QPC, 1987.

Quality Function Deployment; A Collection of Presentations and QFD Case Studies. Dearborn, Michigan: ASI Press, 1988.

Ryan, Nancy, Editor. *Taguchi Methods and QFD.* Dearborn, Michigan: ASI Press, 1988.

VIDEOTAPES ON QFD

The following videotapes are produced and sold by ASI.
  Customer-Driven Engineering: Quality Function Deployment
  QFD and the Competitive Challenge
  QFD: The Budd Company Case Study
  QFD: The Kelsey-Hayes Case Study

Index